Nicholas Grossman is Assistant Teaching Professor of International Relations at the University of Illinois and Editor-at-Large of *Arc Digital*. He is an expert on robotics, drones, terrorism, insurgency, and US foreign policy, and his writing has appeared in *Arc*, *National Review*, *CNBC Opinion*, and elsewhere.

"*Drones and Terrorism* provides an important and needed analysis of the role of drones in the conflict between governments and terrorists. It is one of the only books that addresses the use of drones by these non-state actors and how their use will impact the next stage in the evolution of counter-terrorism. Grossman's research and analysis provides important evidence and arguments to the debate about the role of drones in conflict, going beyond the standard tropes of 'should' or 'should not' to what is actually happening and what to expect in the future. This book also advances our understanding of drones by discussing the next stage in the type and nature of drone warfare, and the implications this will have for conflict. His discussion incorporates how the future development of drones, including a swarm approach, will create challenges and opportunities for policy makers. This book provides academics, students, the public, and policy makers with a very well informed understanding of the future of drones and their role in the continuing fight between governments and terrorists, including how governments should respond to the technological developments in the field of drones. *Drones and Terrorism* moves the discussion about the use of drones forward by providing a much needed discussion of how terrorists use drones, the future of drone technology, and what both of these mean for counter-terrorism policies."

**Brian Lai, Associate Professor of International Relations,
the University of Iowa**

NICHOLAS GROSSMAN

DRONES AND TERRORISM

ASYMMETRIC WARFARE AND THE THREAT TO GLOBAL SECURITY

I.B. TAURIS

LONDON · NEW YORK

Published in 2018 by
I.B.Tauris & Co. Ltd
London • New York
www.ibtauris.com

Copyright © 2018 Nicholas Grossman

The right of Nicholas Grossman to be identified as the author of this work has been asserted by the author in accordance with the Copyright, Designs and Patents Act 1988.

Every attempt has been made to gain permission for the use of the images in this book. Any omissions will be rectified in future editions.

References to websites were correct at the time of writing.

ISBN: 978 1 78453 830 9
eISBN: 978 1 78672 348 2
ePDF: 978 1 78673 348 1

A full CIP record for this book is available from the British Library
A full CIP record is available from the Library of Congress

Library of Congress Catalog Card Number: available

Typeset in Stone Serif by OKS Prepress Services, Chennai, India
Printed and bound by TJ International Ltd, Padstow, Cornwall

MIX
Paper from
responsible sources
FSC
www.fsc.org
FSC® C013056

For William Barr, who taught me to love learning,
and for Elliot Grossman, to whom I hope to teach the same.

Contents

Contents

List of Illustrations

GRAPHS

TABLES

Acknowledgments

I've been working on this idea for a while in some form or another, and I'm very grateful to everyone who helped it get here. Thank you Jason Grossman, Daniel Grossman, Shara Dube, Marc Grossman, Tracy Grossman, a lot of Barrs, a lot of Proroks, Jonathan Selter, Ben Wellington, Jacob Aronson, Lucas McLendon, Michael Carim, Travis Vogan, Brian Lai, Rene Rocha, and many others for talking with me about drones and/or terrorism. George Quester, Shibley Telhami, Paul Huth, Piotr Swistak, Bill Nolte, Keith Olson, and Doug Dion for commenting on earlier drafts. Chris Krugenberg and Grayson Scogin for helping with the research. The Fall 2016 Senior Seminar for work-shopping that sentence (it opens Chapter 2). Tomasz Hoskins, Arub Ahmed, and everyone at I.B.Tauris for making the book a reality. And thank you Alyssa Prorok, for everything.

Introduction

As Iraqi, Kurdish and American forces advanced on Mosul in January 2017, trying to retake Iraq's second largest city from ISIS militants, drones attacked them from above. These were not the large unmanned aircraft the United States uses to monitor and strike suspected terrorists, and they were not firing high-tech missiles. They were small, swept wing planes, about six feet wide, that ISIS modified to carry grenades, mortars and other explosives.[1] Coalition soldiers shot down most of the drones before they could drop their makeshift bombs. But not all.

Other drones stayed back, observing. These were quadcopters, which fly and hover using four rotaries, retail for $1,200 or less, and carry cameras. Somewhere in Mosul, ISIS commanders watched the live video feeds, adjusting their orders. The US-backed coalition does not know the location of ISIS snipers or booby-traps, or where fighters hide among civilians, making urban warfare slow and costly. With drones monitoring the coalition's movements, ISIS sets up ambushes and anticipates when to retreat. The insurgents are outgunned, but this information helps them compensate.

On January 21, while Iraqi soldiers and their American advisers fought ISIS in Mosul, two cars exploded in the town of al Bayda in Yemen, 2,000 miles to the south. US drone-fired missiles destroyed both vehicles, killing at least three members of Al Qaeda in the Arabian Peninsula, including field commander Abu Anis al Abi.[2] Most likely, the strikes came from a Reaper, a sleek unmanned airplane, slightly longer than a four-passenger Cessna. Over the last decade, American drones have launched hundreds of similar strikes, killing thousands.

STRATEGY AND ASYMMETRIC WARFARE

This is a book about capabilities and strategy. What can drones do now, what will they be able to do soon, and how do they change the way states and terrorist organizations fight each other? The more we know about that, the better we can develop security strategies, craft laws, and make informed moral judgments as robotic technology becomes increasingly prevalent in warfare, policing, intelligence, business, and entertainment; firing missiles, spying on suspected criminals, delivering packages for Amazon and other companies, dropping contraband into prison yards, providing innovative angles for sports broadcasts, tracking oil spills, and much, much more.

One of the most significant strategic situations of the twenty-first century is asymmetric warfare: any big-against-small armed conflict, such as guerrilla warfare, insurgency, and terrorism. These competitions are especially interesting because they are inherently unfair. By definition, one side of an asymmetric conflict is stronger, controlling more material resources than its opponent. The weaker side, facing an adversary with more soldiers, more advanced equipment, and more money, has to design strategies that work around these disadvantages. This, in turn, creates challenges for the stronger side as it tries to translate its material advantages into victory.

Terrorists know the target and timing of a planned attack, which gives them an informational advantage over counterterrorists, who must martial their resources to protect many locations at once. For example, the Tsarnaev brothers knew they were going to attack the Boston Marathon on April 15, 2013, but the FBI did not. When their homemade pressure cooker bombs exploded at the finish line, creating a chaos of shouting, blood, and lost limbs, the police had to scramble to catch up. The bombings killed three and injured 264, and it took a three-day manhunt to find the perpetrators. Security cameras near the finish line showed two young men carrying backpacks before the bombs exploded, but not after.

Drone-borne terrorism is still mostly hypothetical—the Boston Marathon bombs were too heavy for inexpensive, commercially available models—but the technology has already transformed asymmetric warfare. The most prominent example is the United States' campaign of drone-fired airstrikes against suspected members of Al Qaeda and other terrorist groups. These attacks are extrajudicial, undertaken without a transparent legal process and conducted outside of official warzones, mostly in Pakistan and Yemen. When they go smoothly, like the strikes in al Bayda that killed al Abi, no civilians die. When they do not, the results are tragic. On December 12, 2013, American drones fired four Hellfire missiles at a wedding procession near the city of Rana'a in Yemen, killing 12 and injuring at least 15.[3] Mistakes like this not only kill innocents, they also undermine American strategy by handing terrorists a propaganda victory.

Though attacks garner more attention, drones spend considerably more time gathering information. Many countries use unmanned aircraft of various sizes—from larger than a fighter jet to smaller than a human hand—to conduct surveillance. Some large American drones carry superpowered cameras that can monitor everything in a 100 square kilometer area, automatically tracking suspects' vehicles 20,000 feet below. In Afghanistan, British soldiers used a micro drone called the Black Hornet, a tiny helicopter that looks like a toy but costs $200,000, to fly around corners, over hills, and into buildings, warning them of ambushes. These informational capabilities help counterinsurgents protect large areas, identify enemies, and avoid civilian casualties.

Terrorists and insurgents also use drones to acquire intelligence and attack targets, though the varieties they employ are usually cheaper and less sophisticated. In a 2006 war against Israel, Hezbollah used Iran-built drones to monitor Israeli troops as they advanced into southern Lebanon. Most were older models; smaller unmanned propeller planes that look like something a child might ride in an aviation museum. However, by 2015, Iran provided Hezbollah with larger, more advanced designs—similar to the streamlined models flown by the US, UK and Israel—which the group deployed in the Syrian civil war.

WHAT'S A DRONE?

More formally known as unmanned aerial vehicles (UAVs), unmanned aircraft systems (UASs), or remotely piloted aircraft (RPAs), drones can fly, do not carry a human pilot and, unlike rockets and missiles, can execute controlled landings and undertake repeated flights. Depending on the model, they fly using propellers or jet propulsion like airplanes, one or more rotaries like helicopters, and even flapping wings like birds or insects. Put more simply, drones are flying robots.

Robots are machines that can perceive their surrounding environment and recognize changes in it, process this information and make decisions in response, and act upon the external environment without constant human direction.[4] By contrast, a computer can processes information and choose among options, but cannot act upon its physical environment, while non-robotic machines (chainsaws, bicycles, lawnmowers, etc.) can affect their environment, but do not make decisions. Robots come in many types. Simple varieties perform repetitive tasks on factory assembly lines, observing an item is in the proper location in front of them, choosing to modify the item instead of doing nothing, and then screwing on a new piece. More complex models can vacuum floors, safely remove explosive devices, navigate highways, map the ocean floor, deliver supplies over harsh terrain, and a lot more.

All technologically advanced militaries now employ robotic systems, which demonstrates widespread belief in their usefulness.[5] As of 2017, only four countries—the US, UK, Israel, and Pakistan—have used armed drones in combat. However, at least six more possess armed UAVs: China, France, Iran, Iraq, Nigeria, and South Africa; and nine are currently developing drones capable of attacking targets: Russia, India, Turkey, Italy, Spain, Sweden, Switzerland, Greece, and Taiwan.[6] Additionally, some militant groups have drone capabilities, including Hezbollah and ISIS (also known as the Islamic State, ISIL, and Daesh). All told, 86 different countries possess UAVs capable of surveillance, with over 60 producing drones domestically, almost all of which also buy UAVs produced elsewhere.[7]

More than 400 companies manufacture UAVs, and while many cater specifically to militaries, at least 228 market drones for individual and commercial use.[8] DJI, the world's largest commercial manufacturer, earned approximately $1.5 billion in revenue in 2016,[9] and investors estimate the company's valuation at $12 billion.[10] Based in Shenzhen, China, DJI makes the Phantom, which is the world's bestselling personal UAV. A small quadcopter—the type of drone ISIS used for battlefield surveillance—the Phantom weighs under three pounds and measures about two feet by two feet. DJI designed the Phantom to carry a camera and hover to facilitate aerial photography. It can fly as high as 6,000 meters, remain in the air for approximately 25 minutes, and retails for $1700 or less, depending on the features.[11] In June 2014, the French national soccer team spotted a Phantom flying over its practice at the World Cup in Brazil,[12] and in the middle of the night in January 2015, a drunken federal employee accidentally crashed one on the White House lawn.

WHERE WE'RE HEADED: THE PRECEDENT OF COMPUTERS

Unmanned technology is already widespread, and will proliferate rapidly in the near future. Though the earliest remote-controlled airplanes appeared before World War I, drones as we understand them today have developed in a pattern similar to computers, and, if the broad outlines of this pattern continue, UAVs will become increasingly commonplace. As the technology spreads and advances, the most advanced militaries will develop increasingly sophisticated unmanned aircraft, less advanced states will acquire additional attack and surveillance capabilities, and it will become progressively easier for terrorists and insurgents to get their own.

At first, the world's most powerful states monopolized information technology, with much of it driven by the military. In 1946, the United States employed the world's first general purpose digital computer to calculate artillery firing tables,[13] and in 1969 launched a computer network called ARPANET, which pioneered the data transmission techniques now utilized by the internet. Computing spread and became increasingly available for civilian use, with personal computers proliferating in the 1980s. The World Wide Web launched in 1991, and, in 1995, commercial service providers brought the internet into private homes.[14] Accompanying this network expansion, the web grew exponentially, from 10,000 pages in 1995 to over 30 million in 2000, and more than 1 trillion unique web addresses by 2010.[15]

Now, in the second decade of the twenty-first century, almost everyone in economically developed countries, and many in the developing world, has access to computers, the internet, and cell phones. With smart phones, average citizens carry portable computers more powerful and with access to far more information than military computers from the 1970s or desktops from the 1980s. Robotics technology is following a similar pattern, with access progressively spreading from governments to large corporations to individuals.

Whereas information technology originally enhanced the capabilities of militaries and research facilities, as computers spread and gained access to the internet terrorist and insurgent groups found ways to utilize them to their advantage. By 2005, every organization on the US State Department's list of identified terrorist groups had a presence on the web, with at least 4,300 separate sites dedicated to the groups or their supporters.[16] In the 2010s, transnational networks, such as ISIS and Al Qaeda, post videos online and maintain a substantial presence on social media. This enhances their ability to recruit, fundraise, spread propaganda, strategize, and share information, from expressions of solidarity to bomb-making techniques.

Computers and cell phones played a significant role in various post-Cold War asymmetric conflicts and terrorist attacks. For example, Hezbollah used a private cell phone network to share intelligence about enemy troop movements in its 2006 war with Israel. Smart phones were essential to Lashkar e Taiba's attack on Mumbai in 2008, as the attackers used their phones' GPS to reach Mumbai by boat, studied online maps to plan their coordinated assault, and monitored news websites during the operation to gather intelligence on each other's activities and the response of Indian security services.

Perhaps most notably, the internet magnifies the self-starter problem, enabling disaffected individuals from many countries to see themselves as part of a global movement. Jihadist sympathizers—from the British-born doctor of Iraqi decent who attacked the Glasgow airport in 2007, to the married couple, one born in America the other in Pakistan, who attacked the Inland Regional Center in San Bernardino, California in 2015—saw themselves as activists fighting for a similar cause; thinking globally but acting locally. It is difficult to imagine this loosely connected transnational network existing without the internet. Since many self-starters either learned how to build explosives from information acquired online, or made contact with terrorist groups over the internet who later supplied them with weapons and tactics, the ubiquity of information technology both connected them to a cause and enhanced their ability to inflict damage.

The spread of robotics technology will probably repeat this pattern, at least along the basic outlines: first enhancing the military capabilities of the wealthiest governments, then assisting with military and non-military tasks of smaller states and sub-national governments along with the commercial efforts of larger corporations, eventually achieving widespread use by individuals. Much as governments and corporations control the world's most powerful supercomputers, these large organizations will likely control the world's largest and most advanced robots. However, the spread of robotics technology will enable terrorists to acquire the cheaper, commercially available versions, and put them to use.

Moving forward, robotics, like information technology before it, will significantly alter asymmetric warfare. Unmanned aircraft systems will enhance the capabilities of states, but their monopoly on the technology will continue to fade as insurgents make use of robots as well. Small unmanned aerial vehicles designed for militaries have already proven useful as information-gathering platforms for state-sponsored networks, namely Hezbollah, and adapted commercially available versions are following suit.

Additionally, as privately controlled flying robots become increasingly commonplace, from UAVs designed for aerial photography to drones delivering food or packages, the chances increase that one will be utilized in a terrorist attack. For example, it is easier for a police officer to stop someone on the ground from taking pictures of a bridge—or another location that bans photography to hinder terrorists' ability to case a target—than prevent a flying robot from doing so. Or someone could load a drone with explosives and fly it into a crowd. Commercially available models are not large enough to take down a building, but they could do some damage. UAVs designed for crop dusting could disperse a chemical agent, and even smaller drones could carry a grenade-sized charge.

The development and proliferation of drones and other robotic systems in the mid-2010s appears analogous to computers' 1980s PC stage. Many countries already allow private and commercial drone flight, and are developing regulatory guidelines to manage the flood of UAVs entering their airspace. And then, in a decade or two, drones' PC stage will lead into a smart phone stage, and UAVs and other robotic systems will be everywhere.

CHAPTER 1

How to Fight an Unfair War

We are living in an age of asymmetric warfare. Nuclear weapons and mutual assured destruction, economic interdependence, and other factors have kept the world's most powerful countries from going to war with each other. With greater resources and more advanced technology, they have imposed their will on weaker states. Since the end of the Cold War, their most formidable opponents in armed conflict have been terrorists and insurgents.

In the twenty-first century, the United States suffered the largest mainland attack by a foreign entity since the War of 1812 at the hands of a transnational terrorist group rather than a state; easily defeated two governments in Afghanistan and Iraq, but had difficulty suppressing the subsequent insurgencies; and engaged in a global campaign against Al Qaeda, ISIS and their affiliates. The "War on Terror" is more than 15 years old, and shows no sign of ending.

Other nuclear powers' post-Cold War experience is remarkably similar. Russia, China, the UK, France, India, Pakistan, and Israel have avoided military confrontation with symmetric adversaries, defeated or coerced weaker states, engaged in difficult conflicts against insurgencies, and lost civilians to terrorist attacks.

Russia easily won a war against Georgia in 2008 and took Crimea from Ukraine in 2014, but has not defeated insurgencies in the Caucuses, and lost 186 children in the September 2004 Beslan school massacre.[1] The UK quickly deposed Saddam Hussein's government in 2003 as part of the American-led coalition in Iraq, but faced considerably more military casualties in a six-year fight against the Iraqi insurgency and over ten years fighting insurgents in Afghanistan, while losing more civilians in a terrorist attack on

London's transportation system in 2005 than in any international conflict since World War II.

India mostly avoided violent confrontation with its longstanding rival Pakistan, losing more soldiers to insurgents in Kashmir, while an attack on Mumbai by Lashkar e Taiba in November 2008 killed 164 and injured 308.[2] Similarly, Pakistan mostly avoided confrontation with India while fighting ongoing conflicts with various militant groups in the Federally Administered Tribal Areas along its northwest border with Afghanistan. For comparison, Pakistan lost 1,174 soldiers in its most recent major engagement with India, the 1999 Kargil War,[3] but in the twenty-first century suffered over 6,000 military fatalities and over 21,000 civilian deaths fighting terrorist and insurgent networks.[4]

Israel cooperated with former adversary Egypt to blockade the Gaza Strip, destroyed a nuclear reactor in Syria with no repercussions,[5] and avoided direct confrontation with Iran while carrying out covert action aimed at delaying Iran's nuclear program.[6] However, Israel failed to achieve its goals in a war against Hezbollah in the summer of 2006; and in three Gaza wars from 2008–2014, Israel lost 79 soldiers and killed thousands of Palestinians, without ever achieving decisive victory. Since 2001, Hamas and other Gaza-based Palestinian resistance networks have fired thousands of rockets fired into Israel, killing 28 civilians, including four children.[7]

The quantity and quality of personnel and equipment are probably the most important factors in warfare. However, a brief glance at history shows they are not the only factors that matter. If they were, the stronger side would always win.

GUERRILLA WARFARE, INSURGENCY, AND TERRORISM

Vietnamese general Vo Nguyen Giap defined guerrilla warfare as the strategy "of the people of a weak and badly equipped country who stand up against an aggressive army which possesses better equipment and technique."[8] Echoing earlier guerrilla theorists, Al Qaeda strategist Abd Aziz al Muqrin writes insurgency "is a war waged by a poor and weak party using the simplest methods and the cheapest means against a strong opponent who has a superiority in arms and equipment."[9] Given their material disadvantage, guerrillas must adopt a long-term view of the conflict, aiming to gradually defeat their enemies both politically and militarily.

"Another fundamental characteristic of the guerrilla soldier" Che Guevara writes, "is his flexibility, his ability to adapt himself to all circumstances, and convert to his service all the accidents of the action."[10]

Guerrilla warfare is based on counterstrategy, creatively employing whatever resources become available, acting underhanded, sneaky, and generally fighting dirty. Guerrillas do not wear uniforms or announce allegiances, allowing them to blend in with the civilian population and hide from retaliation, which Mao Zedong compared to fish swimming in a sea of people.[11] They embrace surprise, sabotage, and assassination, designing hit-and-run raids to exploit enemy weaknesses. While traditional armies aim to capture and hold territory, guerrillas move constantly, harassing the enemy wherever possible.

To succeed in asymmetric warfare, writes international security scholar Andrew Mack, insurgents must first "refuse to confront the enemy on his own terms."[12] In symmetric conflicts, adversaries possess comparable levels of resources and similar military technologies. Though not perfectly equal, they have the resources for a fair fight—for example, in World War I, both sides had machines guns and artillery—which means they rely on superior mobilization, discipline, maneuver, and luck to succeed. By contrast, in asymmetric warfare, direct combat plays to the advantage of the more advanced military. In the late 1800s, when African forces lined up on a battlefield against European colonizers armed with machine guns, the natives got slaughtered.[13]

EVENING THE ODDS

The first step in winning a war against a more powerful enemy is avoiding direct confrontation. Then, having ensured survival, successful insurgents follow one of two main strategies:

(1) Acquire more power while wearing down the enemy's capacity.
(2) Impose military, economic, and political costs until the enemy abandons its military campaign, withdraws forces, or alters a particular behavior.

Domestic conflicts typically classified as revolutions, civil wars, or guerrilla insurrections fit the first form, as revolutionaries intend to take over the state, like the Bolsheviks, or become the dominant governing force of a given geographical area, like the American Confederacy. These asymmetric conflicts end when the challenger to the government becomes the more powerful side and can use the state apparatus to crush opponents, or achieves relative symmetry and can fight with conventional military strategies—essentially, get strong enough to make it a fair fight, or so strong that it is unfair for the other side.

However, many terrorist and insurgent groups cannot plausibly become as powerful as their opponents, and must rely on the second strategy of compellence.[14] This especially applies to nationalist insurgencies, like Vietnam or Iraq, that seek the withdrawal of foreign forces or a decrease of foreign influence, rather than control of the foreign state's territory. Additionally, the second main strategy describes the early asymmetric phases of domestic conflicts that finish as symmetric, such as Mao and Che's communist revolutions. The greater the material disadvantage, the more asymmetric the conflict, and the more restricted insurgents become to strategies that avoid the enemy's strengths.

When, say, the IRA fights the United Kingdom, or Al Qaeda fights the United States, they will always be outgunned. But weapons are not the only factors in asymmetric warfare. Discussing guerrilla strategy, Mao argued "the enemy has advantages only in one respect ... but shortcomings in all others," while insurgents "have shortcomings in only one respect but advantages in all others."[15] This implies insurgents can overcome their resource disadvantage by exploiting non-material asymmetries. For example, Palestinian groups are much less powerful than Israel, which means when the two fight they suffer more casualties than they inflict. But that creates an opportunity for the Palestinians to highlight their underdog status to generate international sympathy, increasing political pressure on Israel to accept a ceasefire.

Resolve: Who Wants it More?

Security scholar Andrew Mack proposed that the weaker side has a greater interest in the conflict. The stakes in asymmetric warfare, he argues, are inherently higher for insurgents because the price of their defeat is the loss of independence or total destruction. The more powerful side wants to win, of course, but does not face an existential threat. When survival is at stake, as in the symmetric World Wars, or domestic revolutions, the war effort takes "automatic primacy above all other goals."[16] However, in international asymmetric conflicts, such as Vietnam, a powerful country like the United States has many interests besides the war, which allows for internal debates over the ideal allocation of resources, creating the political conditions that could lead to withdrawal.

This dynamic means state militaries often face greater political constraints when fighting insurgents than when fighting each other. As Mack writes, "when the survival of the nation is not directly threatened, and when the obvious asymmetry in conventional military power bestows an underdog status on the insurgent side, the morality of the war is more easily questioned."[17] As with the United States in Vietnam, or France in Algeria, domestic and international opposition to the war will grow due to moral outrage over the death and destruction caused by a powerful state asserting a less-than-vital interest. By contrast, when survival is at risk, as for the Allies in WWII, "the propensity to question and protest the morality of the means used to defeat the enemy is markedly attenuated."[18]

By avoiding direct combat, where material advantage could prove decisive, guerrillas can force their opponents into a "protracted war."[19] According to Mao, denying the enemy victory extends the conflict, and creates a situation in which insurgents can slowly bleed powerful armies, imposing costs that weaken resolve. Given enough material advantage, a state will win any contest of force, which means insurgents can win only if they can make the conflict a contest of will. As Mack notes, in cases where the weaker side achieves its goal of political independence, such as Vietnam or Algeria, "success for the insurgents arose not from a military victory on the ground—though military successes may have been a contributory cause—but rather from the progressive attrition of their opponents' political capacity to wage war."[20] The Algerian and Vietnam wars ended not because France and America lacked the military capacity to continue fighting, but because they decided the fight was no longer worth it.

Expectations: Who Thinks They Should Win?
States' material advantage leads them to expect rapid, low-cost victory. Insurgents, on the other hand, face an overwhelmingly powerful foe, and do not expect the conflict to be easy or cheap. As conflicts drag on, the resource toll grows, creating political debates in states with multiple interests, or as Mao put it, "contradictions within the enemy's camp."[21] The more a state

expects a quick and easy victory, the more protracted conflicts and their accompanying resource drain lead to anti-war movements, and arguments to shift resources to other priorities. This applies to autocracies—such as the Soviet Union in Afghanistan—as well as democracies, albeit among a different group of decision makers.

It also disrupts traditional war assessment metrics. Expecting an easy victory, powerful states are particularly affected when they lose soldiers, and less encouraged by enemy casualties. Expecting safety at home, as well as in embassies and bases, they face large political costs from the deaths of civilians or off-duty soldiers. The United States and France, for example, withdrew from Lebanon a few months after the 1983 Beirut barracks bombings killed 241 American and 58 French peacekeepers, determining the mission was not worth the cost.[22]

By contrast, insurgents fighting states expect to lose most battles and suffer greatly in pursuit of victory, and therefore consider individual casualties less costly. Small victories greatly encourage the weaker side, while any developments that do not portend decisive victory discourage the stronger side. As a result, as Henry Kissinger noted, "the guerrilla wins if he does not lose," while the conventional state army "loses if it does not win."[23]

Organization: Large or Agile?
To utilize large amounts of resources, states have adopted bureaucratic institutions. Bureaucratic organizations are based on centralized power and clearly defined positions arranged in a top-down structure (frequently depicted as an inverted tree). The positions and the structure outlast any individual members, who can be replaced with another person fulfilling a similar function. All states utilize this structure to organize their governments and militaries, as do all large businesses, with the most powerful states possessing the largest bureaucracies.

By contrast, many terrorist and insurgent groups have adopted a networked structure. Networks are organized based on nodes (i.e. individuals), and structured by the connections between them. They are "bound together by shared values, a common discourse," and an "exchange of information and services."[24] Networks are more fluid than bureaucracies, changing as individuals leave or join and as the relationships between the members evolve. Insurgencies, transnational terrorist organizations, drug cartels, political activists and many smaller businesses organize as networks. This institutional form is less able to concentrate resources or coordinate actions than bureaucracy, but more capable of changing rapidly and more open to individual initiative. In other words, compared to bureaucracies, networks have less material power, but greater agility.

Terrorists are outgunned by the states they are fighting and, to have a chance at winning, they must minimize their vulnerability to counterterrorist operations. Unlike the vertical hierarchies of state armies—private to general—terrorists and insurgent groups tend to organize as horizontal networks to avoid decisive counterattacks or decapitation strikes. As a result, they are less attached to specific territory, fleeing areas where their enemy is strong, only to regroup and attack elsewhere. Members tend to avoid uniforms or other readily identifiable characteristics, helping them blend in among local populations. Terrorist networks keep the location—or sometimes even the identity—of their leaders hidden, denying their opponents a clear target.

In addition to enhancing the prospects of survival, a networked organization enhances terrorists' ability to surprise opponents with unanticipated attacks and fluid strategies. Networks can employ "idiosyncratic approaches" due to their "cellular and compartmented nature."[25] In *Networks and Netwars*, RAND scholars John Arquilla and David Ronfeldt argue this looser organizational form grants networks a "capacity for swarming,"[26] in which they attack unexpectedly, disperse, and later reform to attack in a different manner. This poses a difficulty for militaries and security services accustomed to fighting bureaucratic opponents, who typically utilize consistent strategies, tend not to hide their members' identities, and operate out of fixed locations.

Unlike bureaucratic state militaries, terrorist and insurgent networks cannot wield vast amounts of material power, but can adapt quickly to changing circumstances. For example, as US Army General Montgomery Meigs argues, the threat Al Qaeda poses to the United States "derives from its ability to change its operational system at will in response to the methods needed to approach and attack each new target."[27] Furthermore, terrorist groups grant greater operational independence to sub-units, decreasing the value of individual captures, and creating the possible threat of sleeper cells.

However, there are numerous disadvantages to networked organizational forms, primarily an inability to exercise concentrated power.[28] Decentralization limits strategic coordination by decreasing the reliability of communications and the efficiency of information sharing. Individual members of a terrorist group may have the same ideology or broad, long-term goals, but disagree on who, where, and how to attack. "As a result, resources may be used poorly, contradictory tactics selected, and activities carried out that serve parochial short-term interests rather than the larger mission."[29]

For example, Abu Musab al Zarqawi, the leader of Al Qaeda in Iraq, prioritized sectarian war against Shia Muslims, attacking their mosques and sometimes beheading them on camera. By contrast, Osama bin Laden and Ayman al Zawahiri, Al Qaeda's central leaders, wanted to unite Muslims in a war against the United States, Israel, and American-allied Arab governments. In 2005, Zawahiri wrote Zarqawi a letter imploring him to focus his attacks on the United States and its Iraqi partners, arguing that, even though Al Qaeda is Sunni, blowing up Shia mosques and beheading Shia Muslims hurt Al Qaeda's image.[30] Zarqawi ignored the instruction, and Zawahiri could not enforce it. In a bureaucratic state military, an officer that disregarded a superior's request like this would be court-martialed.

The importance of trust and interpersonal connections limits scalability, and subjects larger networks to splintering. Disagreement over targets and tactics led Al Qaeda in Iraq to evolve into ISIS and split from the main Al Qaeda organization. As ISIS took control of territory in Syria and Iraq, it began acting more like a government with fixed locations and a bureaucratic organization. That helped ISIS conduct conventional military operations against Iraqi and Kurdish forces, but also provided targets for the American, British, French, and Russian militaries and their local allies. However, ISIS' global organization continues using a loose, networked structure—more of a movement than a single terrorist group—which makes it difficult for governments in the Middle East, North Africa, Europe, North America, and elsewhere to prevent ISIS members and sympathizers from successfully executing attacks.

Responsibility: Protect Everywhere or Disrupt Anywhere?
The less powerful an organization, the fewer its responsibilities to non-combatants. "The insurgent is fluid," writes French counter-insurgency expert David Galula, "because he has neither responsi-bility nor concrete assets; the counterinsurgent is rigid because he has both."[31] Compared to governments, terrorist and insurgent groups are less concerned with maintaining infrastructure, protecting

civilians, managing an economy, and honoring international agreements. They depend on commercially available products, makeshift workshops, the black market and theft for military supplies, rather than an industrial base or international trade. To the extent they provide government-like functions, insurgents are exceeding expectations.

States, by contrast, have greater responsibilities to their populations. People are a primary source of national power, and governments must provide security and basic services or the public will reject them. When non-combatants have the option of assisting or joining an insurgency, a state's need to live up to its governing responsibilities increases; not necessarily because the insurgency is better at providing security and basic services, but because it stands in opposition to the failing government. The state's failure to meet its responsibilities undermines popular support, which reduces the state's access to resources and intelligence, thereby granting the resistance an advantage.

As a result, insurgent violence can be primarily disruptive. Insurgents can hurt states by destroying infrastructure or denying civilians a sense of security, while governments need to protect all major assets at once, requiring far more resources. For example, when Iraqi insurgents set off a bomb in Baghdad, it shows Iraqi civilians the government cannot protect them, and some turn to sectarian militias in response.

A disruptive strategy is attractive to materially disadvantaged combatants because, as Galula argues, "disorder ... is cheap to create and very costly to prevent."[32] By sowing disorder, insurgents force states to devote more resources towards guarding against attacks, increasing the material and political costs of the conflict. This suggests a refinement of Kissinger's maxim: the guerrilla wins if he disrupts the state's ability to function normally, while the state wins only when it eliminates the guerrillas' capacity for disruption.

Information: Who Are the Terrorists and What Are They Planning?

A networked organization is particularly advantageous to terrorist groups because it capitalizes on their informational advantage. Terrorists possess specific information regarding group membership, the allegiance of local non-combatants, and the timing and location of idiosyncratic attacks. The aim of a terrorist group, therefore, is to keep this information hidden from its enemy; and a networked organization compartmentalizes the information so that revelation of a given operation or identity does not compromise the entire group.

Powerful states, by contrast, possess an immense amount of general information. From this general information, they try to identify enemies and anticipate attacks. However, even if they find the needle in a haystack and figure out an attack in advance, they might not move the relevant information through bureaucratic channels in time to act. As the 9/11 Commission discovered, Al Qaeda operatives tripped enough red flags that the September 11th plot could have unraveled, but information did not move from local to federal law enforcement, or between federal agencies, smoothly enough for the government to put the pieces together.[33]

"Identifying the adversary" is straightforward in symmetric wars—with the notable exception of spies and clandestine activity—but not in asymmetric conflicts.[34] Sometimes, counterinsurgents are not even sure who they are fighting. This inability to identify terrorists or insurgents sometimes leads states to employ indiscriminate violence, in the hope of killing combatants along with civilians, or intimidating them into switching their allegiance and providing better information;[35] though this often backfires by galvanizing opposition.[36]

Popular support plays a crucial role in wars between governments and insurgents. Among military and academic scholars, there is a virtual consensus that terrorists and insurgents utilize violence to "alter the attitudes and behavior of multiple audiences."[37] Terrorist groups cannot hope to defeat state armies in direct combat—ISIS will never be able to invade and occupy Paris or Washington DC—and have to design strategies to undermine political support for the conflict among the state's decision makers. To survive, prolong the conflict, and advance their goals, they require some local and international legitimacy, which helps ISIS or Al Qaeda acquire the sanctuary, financial support, freedom of movement, and steady stream of recruits they need to counter Western and Middle Eastern governments' material superiority.

Insurgents usually have greater knowledge of local preferences and forms of communication, and can exploit this information to frame foreign opponents as exploitative and imperialistic. Some organizations, like Hamas or Hezbollah, further enhance their domestic legitimacy by providing social services.[38] In general, states will have greater access to terrorists' private information if local populations consider the terrorists' actions illegitimate. If terrorists use coercion to garner popular support, the state will be unable to counter this intimidation unless civilians believe the state wants to protect them.

Militant groups can enhance their legitimacy by actively promoting their political position and using local knowledge to highlight the most brutal of their opponents' actions. Hamas, for example, often releases photographs or video of Palestinians killed by the Israeli military. Critics allege Hamas deliberately places civilians in harm's way to create these photo ops. However, while Israel possess greater material resources to communicate intentions and spin events, it still cannot delegitimize Hamas among the local population and international sympathizers.

The strategic importance of legitimacy enforces the notion that asymmetric wars are fundamentally political contests. In rare cases, powerful militaries can destroy weaker enemies with unlimited violence. For example, in response to the Warsaw Ghetto Uprising, the Nazis surrounded the ghetto and burnt it to the ground, block by block, killing everyone inside.[39] However, most of the time, a combination of infeasibility and restraint prevent annihilation strategies, even when militaries are willing to deploy considerable force. The United States killed hundreds of thousands, possibly millions in Vietnam, and the Soviet Union did the same in Afghanistan, but both superpowers eventually withdrew without achieving their goals.

As these examples show, almost all asymmetric conflicts depend on a battle for popular support. "All insurgencies," notes the latest US Army/Marine Corps Counterinsurgency Field Manual, "even today's highly adaptable strains, remain wars amongst the people."[40] This does not suggest, as US Air Force General Charles Dunlap mockingly argues, "defeating an insurgency is all about winning hearts and minds with teams of anthropologists, propagandists, and civil-affairs officers armed with democracy-in-a-box kits and volley-ball nets."[41] Capturing or killing committed insurgents will always play a prominent role, but "there is a more certain way of eliminating the guerrilla than seeking to hunt him down among the civilians; it is to turn the civilians against him."[42]

If militaries adhere to basic norms and refrain from massacring whole populations, then asymmetric conflict, by its nature, takes place in the political arena. As terrorism scholar Audrey Kurth Cronin demonstrates, "reducing popular support, both active and passive, is an effective means of hastening the demise of some terrorist groups."[43] Defeating insurgents requires considerable political and military efforts, but as US Army General David Petraeus asserts, "there is no military solution to a problem like that in Iraq, to the insurgency."[44]

Most terrorist organizations face an insurmountable resource disadvantage, and must rely on informational strategies. They utilize violence primarily to demonstrate capabilities and resolve, spread fear, embarrass security services, inspire followers, and provoke overreactions.[45] In this sense, terrorism is a strategy of asymmetric warfare that uses violence against non-combatants or civilian infrastructure to disrupt normalcy, creating a larger psychological and political impact on various audiences.[46] As terrorism scholar Brigitte Nacos argues, "unlike common criminals, terrorists have the need to communicate in mind when they plan and stage their violent incidents; terrorists go out of their way in order to provide the mass media with cruel, shocking, and frightening images."[47] These images and the signals they send are far easier to create than suppress or control.

Strategies based in terrorism frequently fail to achieve perpetrators' ambitious long-term goals, and their success should not be exaggerated.[48] Surprise attacks and mass murder of civilians may backfire by motivating opponents or alienating potential allies. For example, September 11th prompted the American-led invasion of Afghanistan that dislodged Al Qaeda from its base of operations, as well as the subsequent global campaign of asset freezes, arrests, and drone strikes that have significantly weakened the larger Al Qaeda network.

ISIS ran into a similar problem. While it successfully took over parts of Syria and Iraq, terrorist attacks against foreign powers— including a bomb that blew up a Russian passenger jet as it took off from Egypt on October 31, 2015 that killed 224,[49] and a coordinated assault against a soccer stadium, restaurants, and a concert hall in Paris on November 13, 2015 that killed 130[50]— led to increased international efforts against the group. By October 2017, ISIS lost virtually all of the territory it captured to internationally supported Iraqi, Syrian, and Kurdish forces.[51]

Al Qaeda and ISIS are outgunned, but by exploiting factors in asymmetric warfare that can favor the weaker side—resolve, expectations, organization, responsibility, and information—both remain active threats to a variety of countries. Other groups with a more local focus, such as Hezbollah, Hamas, and Lashkar e Taiba, use similar strategies to challenge more powerful opponents, much as the IRA did before. However, drones can help states compensate for the aspects of asymmetric warfare that give terrorists and insurgents an advantage.

CHAPTER 2

The Robotics Revolution

Science fiction promised us robots, but reality is flying past expectations. Humanoid androids with personalities, like C3PO, will not be arriving anytime soon—probably—but a surprising variety of shapes and sizes are already here. Some roll around on treads, with a single robotic arm capable of handling hazardous material, while others walk on four legs, able to recover their balance even after slipping on ice. A robot the size and shape of a softball rolls around, providing operators with a 360-degree image. Large UAVs fly over battlefields, while explosives-laden two-foot-long drones pop out of pneumatic tubes, ready to crash into targets. Multicopters with three, four, or eight rotaries hover and record video, soon to be joined by miniature drones with flapping wings that look like insects.

Many of these automated systems help the world's most powerful states counter the asymmetric threat from terrorist and insurgent networks. Advancements in robotics and information processing have the potential to help states reduce their disadvantages regarding information, resolve, and responsibility. Using robots, advanced militaries can gather and process more information, risk fewer lives, and protect more locations than with human soldiers alone.

COUNTERINSURGENCY VS. COUNTERTERRORISM

In asymmetric warfare, strategies for the stronger side fall into two categories: counterinsurgency (COIN) and counterterrorism (CT). Counterterrorism relies on hunting down enemies. The military and intelligence agencies try to identify active members of a terrorist or insurgent group and kill or capture them. Drones play a major role in the twenty-first-century version of this strategy, locating and following targets with cameras and other sensors, and attacking them with missiles.

Critics deride CT as "Whac-A-Mole," in that the counterterrorists kill enemy fighters in one location, only for new enemies to pop up elsewhere. This criticism applies in the short term, as terrorists and insurgents often avoid areas with a large counterterrorist presence, and convince local civilians not to provide counter-terrorists with intelligence by threatening to hurt them once CT forces leave. More importantly, in the longer term, anger over drone strikes, ground raids and other CT activity aids terrorist recruitment, especially when it results in civilian deaths. As a result, CT could effectively create as many, or even more terrorists than it kills. However, as discussed in Chapter 3, evidence indicates CT can weaken terrorist and insurgent groups' capabilities by eliminating leaders and other valuable members, such as bomb makers.

Counterinsurgency incorporates some of the enemy-hunting elements of counterterrorism, but aims for a larger, more sustainable solution by focusing on protecting the population and strengthening local allies. COIN has three steps: clear, hold, and build.[1] First, military forces clear insurgents out of a given area. Then they hold it, providing security and ensuring insurgents do not return. This sends a signal to the local population that counterinsurgents will protect them, which makes civilians more willing to provide intelligence about insurgents' identities and locations. For example, Al Qaeda in Iraq intimidated bystanders into silence by torturing, maiming, and killing family members of informants, but a sustained counter-insurgent presence in some locations made Iraqis less afraid of cooperating with the United States.

Additionally, in sectarian conflicts, protecting the population helps break the cycle of revenge killing, giving political leaders breathing room to negotiate. Counterinsurgents then work to build up the cleared location, cultivating relationships with locals, helping them improve infrastructure, and training local security forces. This helps convince the population to prefer the counterinsurgents to the insurgents, and provides an exit strategy in which the occupying military hands security responsibility to friendly local forces.

COIN is more resource-intensive than counterterrorism, and places soldiers at greater risk. Holding territory requires more personnel than hunting enemies. To maintain security and earn the population's trust, counterinsurgents need to set up bases and training grounds, and go on patrol, both of which establish patterns and put soldiers in locations where insurgents can attack them. More mechanized forces, which mostly remain in bases and rely on armored personnel carriers and heavier weaponry when venturing out, usually sustain fewer casualties, but interact less with locals, and are therefore less able to cultivate relationships, earn trust and gather human intelligence.[2] In addition, new infrastructure projects provide insurgents with targets they can attack to disrupt normalcy and undermine confidence in counter-insurgents' abilities.

Counterinsurgency takes a long time to succeed, probably a decade or more, because it takes a while to establish trust and build local institutions that can survive the powerful state's withdrawal. The time and commitment necessary to execute this strategy gives insurgents opportunities to impose costs that could convince the foreign power to abandon the effort. For example, British forces and local allies took 12 years to suppress the Malayan National Liberation Army (1948 to 1960).[3] However, even a sustained commitment may be futile, since insurgencies cannot exist without at least a degree of popular support, and counterinsurgents may be unable to change enough minds, or find sufficiently committed and capable local allies.[4] In the aftermath of failed counterinsurgencies, such as the United States efforts in Vietnam, critics often argue the strategy was destined to fail, while advocates say it would have succeeded if the counterinsurgents had more time, devoted more resources, and accepted the necessary risk to their soldiers.

Drones and other robotic systems enhance counterinsurgency strategies by providing the means to sustain the effort at lower costs. By gathering intelligence and striking targets without putting personnel at risk, and protecting soldiers in bases and on patrol, robots can reduce military casualties without sacrificing effectiveness. Drones can also help counterinsurgents monitor and protect more locations at once, reducing insurgents' threat to infrastructure. Building relationships and gathering human intelligence will remain quintessential components of counterinsurgency. However, as more robots take on more COIN tasks, counterinsurgents will be able to carry out their mission while sustaining fewer casualties, under-mining a central component of insurgent strategy.

ROBOTS AND THE UNITED STATES MILITARY

The United States military—the most advanced in the world, with an annual budget greater than the next seven largest spenders combined[5]—has prioritized robotics. Responding to Congressionally mandated austerity and the winding down of wars in Iraq and Afghanistan, the Department of Defense released a document in January 2012 outlining a 22% reduction in total defense expenditures from the 2010 peak.[6] All of this reduction came from personnel and manned systems; the budget protected or increased funding for unmanned platforms. It reduced active Army personnel from 570,000 to 490,000, and active Marine personnel from 202,000 to 182,000, while retiring and divesting planes designed to airlift troops. The 2016 budget continued this trend, further reducing the army to 450,000 active soldiers while maintaining the Marines at 182,000.[7]

Reducing ground capacity and mobility could be expected after mostly concluding two foreign occupations, but the United States reduced manned naval and aerial capacity as well, retiring seven Navy cruisers early, while removing two Littoral Combat Ships and eight Joint High Speed Vessels from future acquisition plans. Additionally, it disestablished six (out of 60) Air Force tactical fighter squadrons. However, the 2012 plan funded the equipment and personnel necessary for 65 Predator and Reaper drone patrols, "with a surge capacity of 85," up from the 2011 total of 61. It also protected or increased the funding for Gray Eagle, the Army's unmanned air system, and "sea-based unmanned intelligence, surveillance and reconnaissance (ISR) systems such as Fire Scout," all in the name of "counter-terrorism operations."[8] The 2016 budget further expands the base number of Predator and Reaper combat air patrols to 76.[9] This is consistent with personnel training over recent years. Since 2009, the Air Force has trained more pilots to fly unmanned aircraft than manned fighters and bombers combined.[10] DoD's priorities are clear: over the next decade, the United States military will become a more roboticized force.

Advances in robotics, along with developments in computing—namely increased networking, information processing, and cyber capabilities—have the potential to grant the United States military significant advantages. While robots undoubtedly would be useful in the event of a war with China, Iran, or another country, their effect on asymmetric warfare will be more immediately dramatic. Unlike states, terrorist and insurgent groups lack the resources to develop or acquire advanced automated systems; and innovations in unmanned technology have already made significant contributions to America's counterinsurgencies in Iraq and Afghanistan, and in the global conflict against Al Qaeda.

PUBLIC OPINION

In the twenty-first century, robots have taken on more combat-related tasks, including some of the most dangerous. With mobile machines of various shapes and sizes turning corners and entering rooms ahead of soldiers, removing wounded troops from combat zones, and searching roads for explosives ahead of human-carrying vehicles, powerful militaries can undertake risky missions with less risk to soldiers' safety. Fewer casualties decreases a major source of the political costs of war, which undermines a primary insurgent strategy in which protracted war and steadily mounting costs create political disputes that eventually lead foreign occupiers to abandon the conflict. For example, the American anti-war movement in response to the protracted conflict in Vietnam would have been weaker if Americans were not coming home in body bags, with robots reducing the need for a draft.

In the United States, as with other countries, public support for a given conflict tends to decrease as casualties rise, with the notable exception of wars against perceived existential threats, such as World War II.[11] Along these lines, it is unsurprising that approval ratings for the war in Afghanistan have steadily decreased, from 90% of Americans approving at the start in 2001, to slightly over half of those polled approving in the mid-2000s, to only 36% approving in 2011.[12] Meanwhile, American fatalities in Afghanistan rose from an annual average of 50 from 2002–2004, to 104 from 2005–2007, and then as high as 499 in 2010 and 418 in 2011.[13] Though American casualties declined thereafter as the United States scaled back its presence, with 127 in 2013 and only 55 deaths in 2014, American public opposition remains high. While in 2002, Gallup found that fewer than one in ten Americans believed "the United States made a mistake in sending military forces to Afghanistan," in February 2014 a majority (49%–48%) believed it, while 42% called the original invasion a mistake in a June 2015 poll, even though most American forces withdrew in December 2014.[14]

By contrast, approval of drone strikes has remained high among Americans. A Washington Post–ABC News poll conducted in February 2012 found 83% of Americans approve of "the use of unmanned 'drone' aircraft against terrorist suspects overseas."[15] The "Global Attitudes Survey," released by the Pew Global Attitudes Project in June 2012, found majorities from most countries disapproving of American drone strikes, with the notable exception of the United States, where 62% of respondents approved and only 28% disapproved.[16]

Unlike many issues in American politics, support for the drone campaign does not differ much based on party affiliation. An Associated Press–GfK poll conducted in April 2015 found nearly 60% of Democrats and over 70% of Republicans support drone strikes against members of terrorist organizations, with only 16% of Democrats and 10% of Republicans opposed.[17] Americans support drone strikes even though they are aware of the potential downsides. In a May 2015 poll, Pew found 58% of Americans approving of the United States "conducting missile strikes from pilotless aircraft called drones to target extremists in countries such as Pakistan, Yemen and Somalia," even though 80% were very or somewhat concerned the strikes "endanger the lives of innocent civilians" and 68% were very or somewhat concerned the attacks "could lead to extremist retaliation."[18]

The gap in these surveys most likely reflects the specific language of each question. The Post-ABC poll specified drone strikes against "terrorist suspects overseas" and the AP–GfK poll asked about targeting "members of terrorist groups," while both Pew surveys did not use the word "terrorist," and the 2015 poll added questions about concerns. Nevertheless, all four polls demonstrate a solid majority of Americans supports the use of drone strikes, and millions of Americans who oppose the war in Afghanistan nevertheless support continuing the campaign of UAV attacks there and elsewhere.

This indicates the American public supports the use of force against suspected members of terrorist and insurgent organizations, except when the effort results in mounting American casualties. Therefore, with an increasingly roboticized military, the United States will be increasingly able to use force abroad without generating much public disapproval at home. This will make America, and other powerful states that utilize drones and other unmanned military platforms, less vulnerable to the Maoist strategy of protracted war, while also raising questions regarding the ease with which governments are willing to use force when they do not have to be as wary of public disapproval.[19] Presidents and prime ministers ordering troops into battle are asking soldiers and their families to sacrifice. But if robots do it instead, there is less of a democratic check on governments' decisions to use force.

GROUND-BASED ROBOTS

Aerial drones get the most publicity, but ground-based robots are revolutionizing twenty-first-century warfare as well. Whether rolling around on wheels or treads, or in a more recent development, walking around on legs, unmanned ground-based systems can enhance the capabilities of soldiers in the field. As of 2017, the United States has developed or acquired robots that can remove wounded troops from battlefields, carry supplies over difficult terrain, detect and remove explosives, shoot firearms with precision, knock mortars and rockets out of the sky, and locate the origin of gunfire. Many of these have already succeeded in active combat theaters.

The United States government has awarded numerous grants to developers of semi-autonomous robots for non-killer tasks. For example, the Battlefield Extraction-Assist Robot, or BEAR, from Vecna Robotics, is designed to carry wounded soldiers to safety without risking others' lives. It can lift up to 500lbs, navigate uneven terrain, climb stairs, and autonomously determine how best to lift objects.[20] The BEAR was invented in 2005, and Vecna received a grant in excess of $1 million from the United States Congress to further its development in 2007. It is undergoing testing at the US Army Infantry Center Maneuver Battle Lab at Fort Benning, where soldiers are growing accustomed to its glove-controller, which recognizes hand gestures, and developing tactics for extracting wounded soldiers in simulated battle conditions.[21]

Another large ground robot partially funded by the United States military is the Legged Squad Support System (LS3) from Boston Dynamics, known as the BigDog, which acts as a robotic pack mule. Unlike the BEAR and most other ground-based robots, which move around on wheels or treads, the BigDog walks on four legs, allowing it to traverse more difficult terrain. With a variety of sensors, a gyroscope, and an on-board computer constantly making adjustments, the robot maintains its balance much like a person or an animal. In demonstration videos, it slips on ice and regains its balance without dropping any of its cargo, all without human assistance.[22] By absorbing the shock of the impact of each leg with the ground, it can recycle some energy from one step to the next, extending operating time between charges.

The LS3 is about 3 feet long, 2.5 feet tall, weighs 240 lbs, and looks eerily like a headless four-legged animal. In separate tests, the BigDog demonstrated it can run 5 mph, climb slopes up to 35 degrees, walk across rubble, through mud, snow, and water, and carry a maximum load of 340 pounds.[23] It also has the ability to follow a human leader without directional input, and, in 2013, Boston Dynamics added a robotic arm capable of lifting (and throwing) heavy objects such as cinder blocks.[24] Funded by the Defense Advanced Research Projects Agency (DARPA), the BigDog began undergoing military tests in 2012, and United States Marines first used a successor LS3 known as Alpha Dog in a real-world situation in the 2014 RIMPAC war games.[25]

In test videos, the LS3 prototype demonstrates that, if knocked down, it can automatically right itself, stand up, and continue walking. Early versions were noisy, but, according to DARPA program manager Lt. Col. Joe Hitt, the latest prototype is "roughly 10 times quieter than when the platform first came online, so squad members can carry on a conversation right next to it, which was difficult before."[26] However, based on the RIMPAC tests, the Marines announced in December 2015 the Alpha Dog was too noisy and would give away its position.[27] Nevertheless, a robotic pack mule would allow soldiers to bring heavier equipment into rougher terrain, and lighten the load carried on their backs, making them simultaneously more mobile and better equipped; and work on new models that utilize quieter electric motors is underway.

BOMB-SNIFFING ROBOTS

Currently, the United States and other advanced militaries make extensive use of smaller, multipurpose robots, like the PackBot by iRobot, which looks like a camera and robotic arm mounted on a platform with treads. The United States deployed over 2,000 PackBots to Iraq and Afghanistan, where they entered buildings or peered around street corners ahead of soldiers, reducing risk to personnel.[28] Weighing approximately 7 to 18 kilograms—give or

FIGURE 2.1   US Marines experiment with an Alpha Dog robot at the 2014 RIMPAC war games.

take, depending on accessories—the PackBot can be carried in a backpack (hence the name).

Most importantly, PackBots can detect and dispose of explosives, especially improvised explosive devices (IEDs). As the following graph shows, of 2,795 International Security Force fatalities in Afghanistan due to hostile action from October 2001 through the end of 2015, 1,401 (50.13%) were due to IEDs. However, the percentage has steadily declined, from a peak of 60.98% in 2009, falling below 50% in 2012, to about 25% in 2014.[29] Part of this decline may be due to shifting insurgent and counterinsurgent tactics, but a significant portion is likely due to deployment of the PackBot 510 EOD model beginning in late 2007. EOD stands for Explosive Ordinance Disposal, and the new model can drag larger objects and lift up to 13.6 kg with its arm in a compact position and 4.5 kg with its arm extended, twice the capability of its predecessors, with a grip three times as strong.[30] Additionally, these robots feature "Fido" sensors that can detect explosive vapors on a level comparable to bomb sniffing dogs.[31]

The decline in the percentage of coalition casualties caused by IEDs coincides with the deployment of thousands of EOD robots, beyond the 2,000-plus PackBots deployed to Afghanistan and Iraq.[32] The larger Talon robot, developed by Foster-Miller and produced by

FIGURE 2.2   American soldiers learning to use a PackBot at Bagram
Air Field in Afghanistan.

QinetiQ, weighs approximately 52 to 71 kilograms, depending on
accessories, and includes chemical, gas, temperature, and radiological
sensors. With its larger size, the Talon is less portable than the
PackBot, but features a more powerful robotic arm, capable of
manipulating heavier objects, dragging up to 113 kg, and lifting up to
34 kg with its arm retracted and 13 kg when extended.[33] Talon's
makers boast of "more than 20,000 successful EOD missions in Iraq
and Afghanistan."[34]

The Talon, PackBot, and other robots with explosive ordinance
disposal capabilities—such as the Wheelbarrow bomb disposal robot,
made by Northrup Grumman primarily for the UK, and the tEODor,
made by Cobham primarily for the Spanish Armed Forces—reduce
insurgents' ability to injure or kill enemy soldiers, thereby weakening
their overall strategy. Less fear of IEDs allows military units to
advance further and faster, while decreasing the rate at which they
accrue costs in prolonged conflicts.

COUNTER ROCKET, ARTILLERY, MORTAR SYSTEMS

In addition to disposing planted explosive devices, automated
systems help protect soldiers against explosive projectiles. After IEDs,

some of the most successful insurgent weapons against American and allied forces in Iraq and Afghanistan have been rockets and mortars. Insurgents occasionally fire these relatively inaccurate projectiles at US bases from nearby residential neighborhoods, thereby discouraging long-range retaliatory fire. The shooters, therefore, often have time to abandon their location before ground forces can respond, making the possibility of retaliation insufficient to deter rocket and mortar fire. To counter this threat, the Army and Marines have employed Counter Rocket, Artillery, and Mortar technology (C-RAM).[35]

In response to an operational needs statement from the Multinational Corps in Iraq, Raytheon adapted its MK15 Phalanx Close-In Weapons System for land use.[36] Since the 1980s, the United States Navy has mounted Phalanxes on ships to protect against anti-ship missiles and aircraft. The system utilizes radar—and, more recently, infrared and electro-optical sensors—to spot incoming projectiles, and then fires up to 4,500 rounds per minute from a swiveling Gatling gun to destroy them before they can reach the ship. Though attached to a fixed position on various vessels, the Phalanx qualifies as a robot because it "autonomously perform[s] its own search, detect, evaluation, track, engage and kill assessment functions."[37]

After tests demonstrating a 60–70% success rate in shooting down incoming mortars, the land-based version known as Centurion was

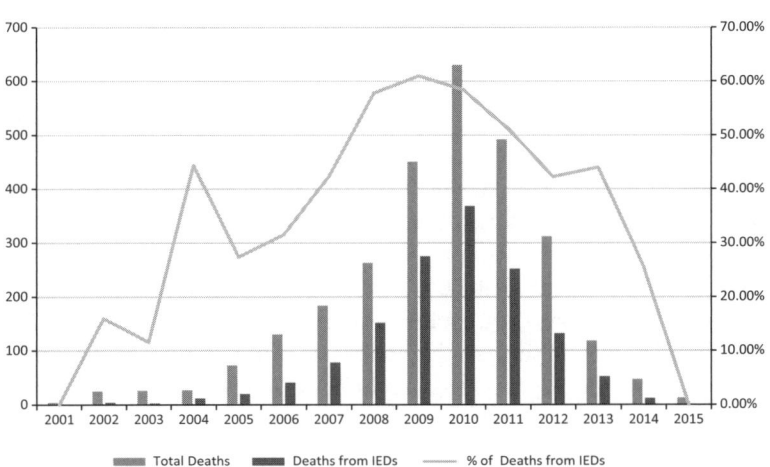

GRAPH 2.1   International Security Force Fatalities: Afghanistan.

first deployed to Iraq in 2005, where the United States installed it at bases and government installations, including the Green Zone and Camp Victory.[38] Unlike the ship-mounted Phalanx, which uses depleted uranium shells, land-based C-RAMs employ incendiary rounds to avoid civilian exposure to radioactive material, and explode mid-air to reduce the risk ammunition that misses the target will harm personnel or civilians.[39] Before deployment, the military required Centurion to demonstrate an ability to neutralize incoming threats while minimizing collateral damage.[40]

While the original Centurion can protect an area of up to 1.2 square kilometers from a fixed position, the latest C-RAM technology aims for greater range, improved tracking, and mobility.[41] In 2010, Raytheon successfully demonstrated a Mobile Land-Based Phalanx Weapon System (MLPWS). This mobile C-RAM system, mounted on the back of a heavy tactical truck, met the 60–70% success rate of its stationary antecedent while maneuvering through 28 miles of paved and off-road conditions.[42] It can provide useful protection against mortars and rockets to mobile convoys, reducing the threat of ambushes, and rapidly send C-RAM defenses to forward positions.

An alternative C-RAM system, which boasts a greater success rate against mortars and rockets, was developed by Rheinmetall for the German military to protect bases in Afghanistan. The Modular Automatic and Network capable Targeting and Interception System, or MANTIS for short, was first deployed in 2011, and includes six 35 mm automatic guns, two sensor units capable of recognizing approaching missiles from 3 km, and a ground-based control unit. Instead of hurling a hail of bullets at incoming projectiles, the system fires air-burst shells that separate into 152 tungsten projectiles, each of which weigh 3.3 grams. The central control unit analyzes information from the sensors to determine the flight path and velocity of incoming targets, and then programs the ammunition using an electronic timer. When the MANTIS' ammunition approaches its target, it bursts into a metal cloud obstructing the projectile's flight path, which increases the chances it will destroy the incoming rocket or mortar compared to other land-based C-RAM systems. The entire process—detection, analysis, counter-fire—takes approximately 4.5 seconds.[43]

To improve range and reduce operating costs, Raytheon is developing a variant of the Phalanx that would use lasers instead of bullets or missiles. By using a focused fiber-optic beam, a C-RAM system could triple the range of earlier models and eliminate the cost of ammunition. However, despite successful tests at shorter distances—in which a laser-based C-RAM destroyed incoming 60 mm mortars at 550 yards—numerous technical problems remain. Lasers require a lot of power to operate, may degrade rapidly when exposed to sand or sea salt, and sometimes have difficulty maintaining full strength in fog or rain. However, an alternative from Boeing, called the High Energy Laser Mobile Demonstrator (HEL-MD), successfully shot down targets in foggy conditions in a 2014 test.[44] Nevertheless, destroying targets with lasers at a greater distance creates additional risk of collateral damage. Unlike incendiary shells, which detonate after a set distance, lasers could go through the target, creating a risk for friendly or civilian aircraft in the area.[45]

Minimizing civilian casualties is essential for C-RAMs to provide strategic value in asymmetric warfare. If insurgents fire mortars or rockets from populated areas, and the C-RAM system knocks the projectiles out of the air without destroying them, they could harm civilians or destroy civilian property. From the insurgents' prospective, both outcomes are strategically beneficial: either the projectile gets through the C-RAM defenses and has the opportunity to strike counterinsurgent soldiers or equipment, or the C-RAM knocks down the projectile in a civilian area, potentially angering the population against the counterinsurgents.

The world's most famous C-RAM system is Israel's Iron Dome, designed by Rafael Advanced Defense Systems. It protects Israeli population centers from rocket and mortar fire, primarily from Hamas and other Palestinian groups based in Gaza. The system detects incoming projectiles and determines their likely destination, firing interceptor missiles at any headed for populated areas and ignoring those that do not pose a threat. Iron Dome first shot down a rocket in April 2011, and by the end of 2014 had shot down over 1,000. The system boasts a 90% success rate, undermining the primary means by which Hamas threatens Israeli civilians.[46]

With rocket fire, terrorists and insurgents exploit states' responsibility to protect their populations. Israelis expect their government to protect everywhere at once, allowing Hamas to disrupt normal life in southern Israel with a few launches. Interceptor missiles are expensive—each costs about $50,000, and Iron Dome fires two at each incoming projectile—while Hamas' homemade Qassam rockets cost under $1,000 and the group does not pay for the more sophisticated Grad rockets it receives from Iran. However, even though Israel spends a lot more shooting down a rocket than Hamas spends firing one, Israel has a lot more money. Nevertheless, a large enough barrage could overwhelm the system, allowing some rockets to get through.

KILLER ROBOTS

Unsurprisingly, in addition to those that protect soldiers and civilians, some modern military robots possess offensive capabilities. The Special Weapons Observation Reconnaissance Detection System, or SWORDS, is a weapons system that can be mounted on a Talon robot. SWORDS replaces Talon's gripping arm with a gun mount that can hold any weapon weighing less than 300 lbs, including an M-16, a .50-caliber machine gun, an antitank rocket launcher, or a 40 mm grenade launcher.[47] In testing, it directly hit bulls-eyes up to 2,000 meters away every time when firing from a stationary position.[48]

The robot achieves greater accuracy than even the best human snipers by eliminating human error. A Talon does not breathe, react to surprises, fear counter-fire, or depend on muscle control, thus providing a more stable platform for weapons than any person could. Furthermore, the SWORDS system matches its zoom lens camera to a weapon's optics, allowing soldiers to see exactly what the weapon is looking at on a monitor, instead of needing to align their eye with the gun sight.[49] Therefore, by using robots instead of human soldiers on the front lines in uncertain, dangerous situations, like urban warfare against an insurgency, militaries not only reduce the risk to their soldiers, but could reduce collateral damage as well.

Futuristic developments over the next decade or two will produce ground-based robots increasingly capable of combat-related tasks traditionally handled by human soldiers. As of 2017, unmanned ground systems can carry supplies, remove wounded soldiers from the battlefield, pick up smaller items, dispose of explosive ordinance, and fire weapons at a target. Soon they will be joined by hunter-killer robots designed to track down and incapacitate human targets.

Adapting the BigDog's ability to walk on legs, Boston Dynamics created an anthropomorphic two-legged robot called PETMAN. It is human-sized and shaped, and can walk, twist, squat, and do push-ups. The company currently sells PETMAN as "an

FIGURE 2.3   An Iron Dome battery firing an interceptor missile at an incoming rocket.

anthropomorphic robot for testing chemical protection clothing."[50] By moving like a person and simulating human physiology, "controlling temperature, humidity and sweating when necessary,"[51] the PETMAN can provide realistic conditions to test suits that protect from poison gas and radiation.

Even though the PETMAN, and Boston Dynamics' follow-up called ATLAS,[52] look eerily like prototypes of the hunter-killer robot from the movie *Terminator*, humanoid robot soldiers are likely still a long ways off.[53] ATLAS can walk on a treadmill and return to its original path when pushed, but it cannot move as smoothly or as quickly as a person. On video, its jumping-jacks and push-ups appear stunted, and it lacks the ability to rapidly switch motions, such as from walking to crawling.[54]

With engineering improvements, a humanoid robot could move faster or more smoothly than the current incarnation of ATLAS, but the lack of high-functioning artificial intelligence means human soldiers are in no danger of being replaced by robots in the near future. For example, DRC-HUBO, designed by a team from the Korea Advanced Institute of Science and Technology (KAIST), won an international robotics competition sponsored by DARPA in June 2015 by autonomously completing eight human-like tasks such as getting out of a vehicle, opening a door and climbing steps.[55] While impressive from an engineering standpoint, DRC-HUBO took 45 minutes to complete a basic obstacle course a child could have handled much more quickly, partially because it took long pauses to figure out how to approach the tasks.[56] As this indicates, no currently designed robot possesses anywhere near the adaptability or decision making capabilities of human beings.

However, the ideal hunter-killer robot to assist human soldiers in a limited task may not be humanoid. In the *Terminator* movies, the robot resembles a person to infiltrate human society and track down a specific target. It looks human to avoid detection, and interacts with people to acquire information. But, unlike an army of evil machines, we have people for that.

Appearing and acting exactly like a real person is far beyond the capabilities of current robotic technology. Human soldiers will remain central to military tasks, especially the population-interaction elements of counterinsurgency, for the foreseeable future. However, a ground-based hunter-killer robot could help human soldiers catch a fleeing suspect.

In pursuit of this goal, DARPA granted a contract to Boston Dynamics to adapt the BigDog into a cheetah-like robot that can track down human prey. Like the BigDog, the Cheetah-bot stands on four legs, but the aim is to create a faster, more agile robot capable of making tight turns so it "can zigzag to chase and evade."[57] It could have non-combat uses as well. For example, DARPA and Boston Dynamics foresee this fast, four-legged robot assisting emergency response teams, reaching victims of a fire or vehicular accident before human first responders. In a 2012 demonstration, the Cheetah-bot reached a maximum speed of 28.3 mph galloping on a treadmill, which is faster than Olympic sprinters.[58]

As ground-based robots increasingly protect soldiers from hostile fire and take on some of the most dangerous combat-related tasks— including battlefield extraction, explosive ordinance disposal, and front-line advancement on enemy positions—insurgents' ability to kill an advanced military's soldiers will likely decrease. With fewer soldiers returning home in body bags, governments would face less domestic political pressure to negotiate unfavorable settlements or abandon protracted conflicts. Additionally, the increased precision of robot-fired weaponry could decrease civilian casualties without reducing military effectiveness. By shooting more accurately and

FIGURE 2.4   Boston Dynamics' ATLAS robot.

never acting out of fear for their own safety, robots like the SWORDS system or a futuristic hunter-killer could decrease the harm to civilians that fuels both insurgents' recruitment efforts and political pressure from human rights groups and anti-war activists. Ground-based robots, therefore, have the potential to help the stronger side win asymmetric conflicts more often.

AERIAL ROBOTS

In addition to ground-based automated systems, the United States military and intelligence community makes extensive use of flying robots. Most drone missions involve reconnaissance, but some UAVs, like the Reaper, use missiles to attack targets on the ground. As of 2017, there are at least 81 different UAV models in use, discontinued, or currently in production.[59]

Unmanned aircraft are almost as old as manned airplanes, with the first attempts at remote-controlled planes coming in World War I. The Hewitt-Sperry Automatic Airplane project aimed to create a "flying bomb," and flew the unmanned N-9 model for the first time in 1917. An explosives-laden unmanned airplane, the N-9 was more of a precursor to cruise missiles than modern UAVs. The earliest unmanned aircraft using a jet engine, the Firebee by the Ryan Aeronautical Company, first flew in 1955, primarily as target practice for aircraft gunners. Later versions, including the Ryan Lightning Bug, were designed for reconnaissance, and the United States first flew Lightning Bugs in August 1964 to gather information over China and Vietnam.[60]

Expanding UAV use in combat, Israel utilized adapted Firebees as decoys to distract Syrian air defenses in the 1973 Yom Kippur War. In 1982, in advance of a strike on Syrian positions in the Bekaa Valley, Israel sent a squadron of UAVs that broadcast signals like regular planes, prompting Syrian anti-aircraft fire. The Israelis then followed with a wave of manned aircraft that destroyed the air defenses using radar frequencies revealed by the anti-aircraft batteries' attacks on the drones.[61]

LARGER DRONES

However, in the post-Cold War world, UAV use has expanded dramatically, most notably for targeted killings. The Predator drone, manufactured by General Atomics Aeronautical Systems, comes in two versions: the RQ-1 for surveillance and reconnaissance, and the

MQ-1, which includes combat capabilities. Operational since 1994, the Predator first flew missions in Bosnia in 1995, in support of forces under the auspices of NATO and the United States.[62] It can fly up to 25,000 feet, reach a maximum speed of 136 mph, and remain in the air up to 24 hours.[63] The original version, used in the former Yugoslavia, was flown remotely by a pilot and sensor operator, sometimes accompanied by payload specialists, sitting in a van near the runway of the drone's operating base. Direct radio signals controlled takeoff and landing, just like a remote-controlled model airplane. Once airborne, communications between UAV and pilot shifted to the military's satellite network, which often caused delays of a few seconds between a pilot's command and the drone's response.[64]

By the beginning of the twenty-first century, improvements in communications technology allowed pilots to fly unmanned aircraft from thousands of miles away, without much noticeable delay. The United States has two main UAV programs operating out of two command centers: CIA pilots fly drones from the agency's head-quarters in Langley, Virginia, near Washington DC, while most of the United States military's armed UAV pilots operate out of Creech Air Force Base in Nevada.[65] Both bases are over 6,000 miles away from Afghanistan, Pakistan, Somalia, or Yemen, where the planes fly and execute various missions, including missile strikes.

The Predator was the first UAV controlled via satellite data link, the first to support manned aircraft with target laser designation, and the first to fire air-to-ground missiles.[66] As a result, it is also the first flying robot in history to kill people both inside and outside of a war zone. In February 2001, an MQ-1 Predator successfully fired a Hellfire-C laser-guided missile in flight tests at Nellis Air Force Base in Nevada,[67] and in November of that year, the United States military used Predators to strike targets during the invasion of Afghanistan.[68] In November 2002, a CIA-controlled Predator destroyed a jeep in Yemen with a Hellfire missile, killing six men, including Ali Qaed Senyan al Harthi, a member of Al Qaeda linked to the bombing of the USS Cole off the coast of Yemen on October 12, 2000.[69]

Since the first attacks in 2001 and 2002, the MQ-1 Predator—and the Predator B, a successor known as the MQ-9 Reaper—have played an increasing role in American counterterrorism and counterinsurgent efforts. The Reaper is a larger, more powerful version of the Predator, specifically designed to strike enemy targets.

FIGURE 2.5   Arming an MQ-1B Predator in Iraq.

According to United States Air Force General T. Michael Moseley, "the Reaper represents a significant evolution in UAV technology and employment. We've moved from using UAVs primarily in intelligence, surveillance and reconnaissance roles before Operation Iraqi Freedom, to a true hunter-killer role with the Reaper."[70] First flown in 2001, the Reaper features a 900-horsepower engine, compared to the original Predator's 119hp, and can carry 15 times the ordinance, fly twice as high and at least twice as fast as the earlier model.[71] General Atomics is developing an even faster version, the Predator C, or Avenger, which first flew in April 2009, uses a turbofan jet engine instead of a propeller, can reach 50,000 feet and carry larger payloads.[72] In October 2016, the Avenger flew its first mission, dropping leaflets over Syria.[73]

From 2001 to 2015, there were 237 "major" United States military UAV crashes—defined as those that destroy the aircraft or cause at least $2 million in damages—and 66% involved Predators or Reapers controlled by the United States Air Force. Twenty crashed in 2015 alone.[74] With the exception of one shot down over Syria by anti-aircraft fire, all 2015 crashes were accidents. Though none hit people, there have been some near misses.

Investigators traced the problem to a faulty starter-generator, but as of yet do not know why it stops working properly. The drones were

FIGURE 2.6   An MQ-9 Reaper flying over Afghanistan.

carrying batteries for emergency backup power, but they only last an hour, forcing pilots to deliberately crash the aircraft in a remote area if they could not get it to an airfield in time. In response, the Air Force added backup generators capable of providing power for up to 10 hours, giving the drones enough time to land. These incidents of mechanical failure indicate the limits of current technology, and place some stress on the operational Predators and Reapers to meet requests for surveillance. However, they have not discouraged the United States or other governments from plans to rely on growing fleets of UAVs.

Though the Israeli government does not officially admit to using armed drones, the first reliable reports of an Israeli drone strike come from October 2004, most likely fired from a Heron, made by Israel Aerospace Industries.[75] Though the Heron looks different from the Predator, the two are of similar size—both about 27 feet long, with wingspans of about 55 feet—and can carry a similar payload: a maximum of 450 lbs for the Predator and 550 lbs for the Heron.[76] The first confirmed Israeli drone strikes killed two members of Islamic Jihad in Gaza on October 24, 2004, and another in Gaza on December 7.[77] Subsequently, Israel fired numerous drone strikes against Palestinian targets in Gaza, as well as against Hezbollah in Lebanon during the Israel–Hezbollah war of 2006, using versions of the Heron, as well as the smaller Hermes UAV made by Elbit Systems. Additionally, Israel used drone strikes to destroy convoys in Sudan suspected of carrying Iranian arms to Hamas in January and February 2009, kill four Sinai-based militants in Egypt in August 2013,[78] and kill two members of a Syrian rebel group called the National Defense Forces in Syria on July 29, 2015.[79]

In May 2008, the UK became the third country to fire weapons from unmanned aircraft. In Afghanistan, Reapers controlled by the Royal Air Force dropped 51 laser-guided 500lb GBU-12 bombs from May 2008 through the end of 2011, and fired 459 Hellfire missiles between May 2008 and November 2014.[80] Against ISIS targets in Iraq and Syria, RAF Reapers fired 545 missiles from September 1, 2014 to September 30, 2016, which according to UK estimates have killed about 1,000 Islamic State fighters.[81] British officials claim "to date there have been no known cases of civilian casualties resulting from UK strikes in Iraq," though critics such as Human Rights Watch express skepticism, noting the British government has not supported its claim with publicly available evidence.[82] Additionally, the UK fired at least one drone strike in Syria in August 2015 that killed two British citizens who were allegedly ISIS operatives.[83]

Much like ground-based robots, aerial drones allow militaries to execute missions at reduced risk to personnel. This leads to fewer soldier deaths, and, correspondingly, less political cost. States do not want unmanned aerial systems to crash or get shot down, because of the cost of the equipment and the risk enemies will learn more about proprietary technologies; but there is no danger a human pilot will be killed or captured and exploited by enemy forces.

Drones aren't cheap, but cost a lot less than manned aircraft. They do not include the equipment necessary to accommodate a person, such as a cockpit, ejection seat and parachute, or air pressure control. For example, each new F/A-18 Hornet Fighter costs $57 million,[84] while the more advanced F-22 costs $143 million,[85] and each F-35 Joint Strike Fighter costs $98–$116 million, depending on specifications.[86] These are the marginal "flyaway" costs of each new unit, and do not reflect sunk costs, such as research and development. Accounting for all expenses, the F-22 cost the United States as much as $678 million per plane.[87] Comparatively, the flyaway cost of each Reaper is a relatively cheap $16.05 million, which includes the ground control equipment, satellite uplink, and the drone itself.[88]

Manned aircraft are not yet obsolete, because they offer superior air-to-air combat capabilities. Unlike fighter jets such as the F-18 or F-22, Predators and Reapers lack the ability to engage in aerial dogfights against enemy planes. The drones are capable of carrying air-to-air missiles, which they could fire at opposing aircraft, but they lack the speed and maneuverability of fighter jets. Reapers use a turboprop engine and cruise at 230 mph, but the F-22's dual turbofan engines enable sustained supersonic flight at over 1,200 mph. Newer, jet-powered UAV models fly faster than Reapers, but still lack the situational awareness to challenge manned fighters.[89]

In December 2002, before the start of the Iraq war, an MQ-1 Predator gathering information over Iraq was fired upon by an Iraqi MiG-25 Foxbat. As the MiG's missile approached the Predator, the drone's pilot launched an air-to-air Stinger missile in response, but did not connect. The Predator was destroyed.[90]

Since then, there have been no reported incidents of UAVs firing upon enemy aircraft. Due to the absence of air threats in the Iraq and Afghanistan wars, and Predators' limited carrying capacity, the United States outfitted its MQs exclusively with ground attack weapons;[91] but in the future, the United States and other advanced nations will likely develop drones capable of aerial combat. A 2009 Air Force study mapping out the future of unmanned aircraft envisions a class of UAVs called "MQ-Mc," which would be capable of any Air Force mission, including dogfighting and nuclear strikes, by 2030.[92] Meanwhile, the Navy is already designing experiments in which two teams, each made up of as many as 50 small "aerial battle bots," will engage each other to develop tactics for unmanned air-to-air combat.[93] Besides costing less than manned aircraft, and eliminating the risk to human pilots, unmanned planes have the potential to be more maneuverable, because human pilots can lose consciousness from the g-force of rapid turns at supersonic speeds.

However, air-to-air combat capabilities are not especially important to asymmetric warfare. Due to their resource advantage, powerful states can easily maintain air superiority over insurgent opponents. The primary threat to their aircraft comes from surface-to-air weapons, such as shoulder-launched missiles, rather than enemy fighter jets. Every fixed-wing or rotary aircraft lost by the United States and allies in Iraq or Afghanistan was due to accident or ground-based fire. Therefore, while drones with the ability to engage in aerial combat would provide strategic advantages in a symmetric conflict between powerful states, UAVs with information-gathering and ground attack capabilities are sufficient for asymmetric conflict against terrorist and insurgent networks.

SMALLER COMBAT DRONES

In addition to large drones, like the Reaper, the United States employs smaller unmanned aerial vehicles that help ground forces attack with greater precision, reducing the risk to soldiers and nearby civilians. For example, the Switchblade, from AeroVironment, is "a Non Line of Sight (NLOS) weapon" measuring only two feet from nose to tail, and weighs a little over two pounds.[94] It launches out of a mortar-like tube, whereupon its wings pop out (like a switchblade knife) and its camera switches on.[95] Alternatively, it can launch from the 70 mm rocket tubes used on army helicopters.[96] Together, the launch tube and drone weigh 5.5 lbs, and one solider can easily carry it.[97] Using a hand-held controller that receives video and GPS coordinates, an operator can guide the Switchblade and then crash it into a target in a kamikaze attack.

This provides troops in the field with a valuable method of attacking distant targets without having to call in airstrikes. The Switchblade utilizes a quiet electric motor, which allows it to sneak up on targets,[98] can remain in the air for 20 to 40 minutes,[99] has an effective range of 10 kilometers, and is capable of suspending its attack sequence and loitering.[100] While Predators and Reapers fire 100-pound Hellfire missiles, or drop 500-pound GPS-guided bombs,[101] the Switchblade carries an explosive comparable to a hand grenade.[102] It therefore causes far less collateral damage to bystanders or property. After successful tests in 2011, the United States Army ordered over 100 Switchblades,[103] awarding AeroVironment with a series of contracts that total $10 million for the drones and associated services, such as training.[104] By September 2014, the Army had increased funding to $63.8 million, with an option to go as high as $102.1 million.[105]

In many ways, the Switchblade represents a culmination of the century-old effort to create a flying bomb. While missiles are capable of quickly traveling great distances, and the modern varieties can shift direction mid-air, they cannot hover or return to base. The Switchblade, however, can loiter and land, giving operators the ability to pause an attack to reconsider, or call it off and reuse the equipment later. Additionally, the cameras allow soldiers to pursue fleeing suspects and confirm a target's identity at close range before initiating the attack sequence.

The Switchblade is ideally suited for urban warfare, as it facilitates attacks against covered positions, and grants soldiers the ability to strike enemies firing from rooftops or windows without destroying entire buildings. Firing mortars, lobbing grenades, or calling in airstrikes risk collateral damage, while advancing on the enemy's position places infantry at risk. Additionally, as soldiers advance, they often utilize cover fire, which could accidentally hit bystanders. The Switchblade, however, can maneuver around objects and strike directly around corners, over walls, at fortified positions or enemies hiding from a soldier's line of sight. This decreases the ability of insurgents to manipulate popular support by operating in populated areas and provoking attacks, because small kamikaze drones like the Switchblade help counterinsurgents respond to enemies firing from covered or hidden positions without risking extensive civilian casualties.

SUPPLY DELIVERY

In addition to unmanned attack aircraft, the United States military has used pilotless helicopters to deliver supplies in combat theaters. Starting in December 2011, two modified K-MAX helicopters, built by Kaman and modified for autonomous flight by Lockheed Martin, began transporting goods to American Marine outposts in Afghanistan.[106] These experimental missions were such a success the Marines extended the program twice beyond its original six-month limit, sending the K-MAX home in 2014 as the United States drew down forces.[107] At the end of 2012, the autonomous helicopter won both Popular Science and Aviation Week's top annual award.[108]

The K-MAX can carry up to 6,000 pounds at sea level, which is more than its empty weight, and more than 4,000 pounds at an altitude of 10,000 feet, attached to a steel cable. With its "four-hook carousel," the helicopter can drop off supplies in multiple locations on one mission.[109] In the first two months, K-MAXs delivered over 100,000 lbs of cargo on over 50 unmanned resupply missions.[110] By the end of their mission, after nearly three years of use, the automated helicopters had transported more than 4.5 million pounds.[111]

Unlike the Predator and other fixed wing UAVs, which are usually piloted by remote control, the unmanned K-MAX often flies autonomously. Originally, the helicopter flew along a pre-programmed course to a forward operating base using GPS coordinates, where a person on the ground directed the drop with a remote control. However, the K-MAX can now deliver cargo without human intervention. Using a beacon approximately the size of a hockey puck to mark the drop point, an unmanned K-MAX autonomously deposits cargo within three meters of its target.[112] By using a variety of sensors, it is able to land or drop off its cargo in total darkness.[113]

Whether partially or completely autonomous, robotic transport helicopters provide two main advantages to counterinsurgents: bypassing land-based supply routes and reducing the risk to helicopter pilots. Ground forces, especially those stationed in remote locations, require a steady supply of food, fuel, ammunition, and replacement parts for equipment. Traditionally, armies convoy materials to forward troops with long, ground-based supply lines, along which vehicles—and, in rougher terrain, pack animals—could be attacked.

Supply line disruption is a common insurgent technique, because convoys are rarely as well armed as combat troops, and regular routes allow those with knowledge of the territory to set up ambushes. Helicopter-based supply lifts reduce the need for ground-based supply lines, which reduces insurgents' ability to ambush convoys, kill personnel, deny materials to troops in the field, and, perhaps most importantly, capture supplies. Switching from ground-based to aerial supply missions would make Che Guevara's favored technique of supplying his forces with captured materiel much more difficult.

Automated helicopters increase the feasibility of an aerial alternative to ground-based supply lines. Like fixed-wing UAVs, the automated K-MAX does not get fatigued or hungry, and can remain in flight longer than manned aircraft. As with other drones, robotic helicopters eliminate any physical risk to human pilots. However, this is arguably more important for helicopters than airplanes, because helos fly lower and slower than planes, making them more vulnerable to ground-based anti-aircraft fire.

As with ground-based robots, unmanned airplanes and helicopters reduce casualties, thereby slowing the rise of war-weariness and the associated political pressure to abandon protracted conflicts.

Additionally, drones' ability to wait longer than manned aircraft before striking reduces the risk of collateral damage, generating less anger among the population. As a result, states face fewer political costs, and insurgents experience less of a boost to recruitment.

SPY DRONES

While drone strikes garner more publicity, most UAV flight time is devoted to gathering information. For example, in Afghanistan from 2009 to 2011, the United States conducted more than four times as many spy sorties as strike missions.[114] Robots do not attack in fundamentally different ways from humans; UAVs fire the same types of missiles as manned aircraft, and ground-based bots fire the same types of weapons humans carry or mount on manned vehicles. However, a robot can gather far more, and more detailed, information than a person by employing daylight cameras, infrared cameras, radar, and other sensors. Furthermore, machines can process more information, more quickly, and from more sources at once. The rapid advancement of robotics and information technology show no signs of slowing, dramatically impacting asymmetric warfare by helping counterterrorists and counterinsurgents overcome asymmetries of information.

While the earliest unmanned aircraft were attempts to create flying bombs, the United States began developing the forerunners of modern UAVs to replace spy planes. In 1960, the Soviet Union shot down an American U-2 over Sverdlovsk (now Yekaterinburg) that was using high-resolution cameras to photograph military installations and other strategically important sites on Soviet territory. The pilot, Francis Gary Powers, managed to eject and parachute down safely, but was captured by Soviet forces, along with the remains of the mostly intact U-2. The incident caused considerable embarrassment for the United States, and led to the release of KGB colonel Vilyam Fisher in a prisoner exchange for Powers. Within days of Powers' capture, the United States launched Red Wagon, a classified UAV program.[115]

American drones—primarily Ryan Lightning Bugs—began flying reconnaissance missions in the 1960s over Vietnam and China. The United States Air Force's 100th Strategic Reconnaissance Wing flew 3,435 UAV missions during the Vietnam War, losing 554 unmanned planes. In Congressional testimony, United States Air

Force General George S. Brown explained the straightforward logic: "The only reason we need [UAVs] is that we don't want to needlessly expend the man in the cockpit."[116]

The modern UAV successor to large, high-endurance spy planes like the U-2 is the RQ-4 Global Hawk made by Northrup Grumman. The RQ designation identifies the Global Hawk as an intelligence-gathering platform, in contrast to the MQ designation identifying the Reaper as a combat system. First tested in June 1999, the Global Hawk can fly extremely high, up to 65,000 feet, and remain in the air for as long as 35 hours. With a maximum speed approaching 400 mph, the Global Hawk can fly 1,200 miles to a target area, observe the area for 24 hours, and then return to base. And it is almost entirely autonomous. Once programmed where to fly and what area to observe, the Global Hawk can autonomously taxi, take off, fly, gather information about the target area, return, and land. Ground-based operators, primarily at Beale Air Force Base in California, monitor the UAV remotely and can redirect the plane or its sensors as they wish.[117]

Twenty-first-century intelligence, surveillance, and reconnaissance (ISR) systems like the Global Hawk utilize a variety of methods to gather information. In addition to high-resolution cameras that provide photographs and video, ISR drones carry infrared sensors, which observe heat, rather than visible light, allowing them to identify hot objects like people, vehicles, anti-aircraft batteries, electricity generators, and computer servers. Synthetic-aperture radar, which spacecraft use to observe the surface of planets and other celestial objects, utilizes the motion of the aircraft and a series of sound waves to provide a detailed map of terrain, including land formations, buildings, and other objects.[118]

Complimenting these are electro-optical sensors, which gather information about a given object by analyzing the spectrum of electromagnetic energy—infrared, visible, and ultraviolet light—it reflects and absorbs. Much as space telescopes can determine the chemical make-up of distant stars by the electromagnetic energy they emit, electro-optical sensors on ISR drones can determine the type and strength of fuel coming out of the back of a missile, as well as distinguish between objects that appear similar to each other in photographs, such as natural foliage and artificial camouflage.[119] Beginning in 2007, some UAV models feature the Airborne Signals Intelligence Payload system (ASIP), which tracks and identifies radar and other types of electronic and communication signals.[120] The

FIGURE 2.7   An RQ-4 Global Hawk undergoing maintenance.

infrared, radar, electro-optical, and electronic signals sensors can gather information day or night, regardless of cloud cover.

Utilizing these sensors in combination, the Global Hawk can conduct a wide-area search observing an entire region, or focus on a single target using its "high-resolution spot mode."[121] In 24 hours, it can image a 40,000 square-mile area, approximately the size of Ohio, and relay this information in near-real time using satellite and ground-based communication systems.[122] Northrup Grumman boasts that Global Hawks logged over 350 hours of flight time in the Iraq War, collecting over 4,800 images, and locating surface-to-air (SAM) missile batteries, SAM transporters, and Iraqi tanks.[123]

Unlike the Global Hawk, the RQ-170 Sentinel from Lockheed Martin features stealth technology, making it better suited for gathering information against targets that possess air defense capabilities. Introduced in 2007, the Sentinel is operated by the Air Force and the CIA, and much of its specifications remain classified. In contrast to other drones, the Sentinel is a flat "flying wing," and looks like a smaller version of the B-2 stealth bomber. Because it utilizes jet propulsion, the Sentinel can fly faster than propeller-powered UAVs like the Predator, and reach heights of 50,000 feet.[124]

Photographs caught the Sentinel flying over Afghanistan in 2007, earning it the nickname "the Beast of Kandahar,"[125] and it played a role in the operation that killed Al Qaeda leader Osama bin Laden. It is widely assumed the Sentinel possesses an array of sensors similar to other ISR drones like the Global Hawk, with the possible addition of nuclear material "sniffing" sensors that can detect radioactive isotopes at a distance.[126] With its role providing real-time battlefield intelligence to ground forces, along with suspected spy missions over Iran and North Korea, this stealthy UAV demonstrates both the rapid advancement of drone technology, and the increasing usefulness of unmanned systems.

AN ALL-SEEING EYE IN THE SKY

In 2010, the United States began outfitting reconnaissance drones with the next generation of ISR cameras, a wide area airborne surveillance system from the Sierra Nevada Corporation. Nicknamed Gorgon Stare, after the unblinking monsters from Greek mythology, the system uses nine electro-optical and infrared cameras to observe up to 100 square kilometers at once.[127] The images have enough detail for the system to send up to 65 different views to different users on tablets or laptops, allowing some users to zoom in on a small section while another simultaneously looks at a wider area. According to Maj. Gen. James O. Poss, the Air Force's Assistant Deputy Chief of Staff for Intelligence, Surveillance and Reconnaissance, "Gorgon Stare will be looking at a whole city, so there will be no way for the adversary to know what we're looking at, and we can see everything."[128] Gorgon Stare thus provides a significant advancement from one-camera systems that could capture video images of a single target, like a building or an intersection.

The Air Force plans to mount Gorgon Stare on Reaper UAVs, and reportedly began using it in a limited capacity in Afghanistan beginning in December 2010.[129] The system weighs 1,100 pounds, and, because of its weight and configuration, flies on Reapers that are not also carrying weapons.[130] The United States has ordered at least eight, at a cost of $17.5 million each.

However, Gorgon Stare disappointed in tests in late 2010 conducted by the 53rd Wing of the Air Combat Command at Eglin Air Force Base, which deemed the system "not operationally effective" and "not operationally suitable."[131] It successfully tracked vehicles, but could not reliably follow smaller objects, most notably people. Gorgon Stare sometimes failed to seamlessly join images from

FIGURE 2.8   An illustration of an RQ-170 Sentinel.

multiple cameras, creating blind spots and leading the system to lose track of objects as they left an individual camera's frame. Limited bandwidth combined with huge amounts of data caused delays in relaying information to the ground. Most egregiously, even when it successfully tracked objects, a software error occasionally generated "a faulty coordinate grid," sending an inaccurate location to operators. This could lead forces acting on the information to lose an object of interest by arriving at an incorrect location, or, disastrously, attack a civilian or friendly target. These problems led one tester to deem Gorgon Stare only "55 to 65 percent reliable,"[132] which is insufficient for regular use in the field, especially regarding information armed forces act upon in real-time.

These difficulties are technical, rather than conceptual, and will almost certainly improve as imaging and data transfer technology continue to progress. A promising alternative is a system based on a single, extremely powerful camera rather than a series of integrated sensors like Gorgon Stare. The Autonomous Real-Time Ground Ubiquitous Surveillance Imaging System (ARGUS-IS), developed by BAE systems, utilizes the world's highest resolution video camera.[133] At 1.8 gigapixels, it can spot a 6-inch object from 17,000 feet in the air.[134] The picture has so much detail that the ARGUS can provide over 60 independent "electronically steerable" windows that zoom in on a component of the larger image. Instead of directing a camera to change its focus, a computer system focuses on an aspect of the recorded image, either providing continuous footage of a fixed area, or automatically keeping a specific target in the window.[135] Therefore, unlike Gorgon Stare, the ARGUS does not lose track of an object as it moves from one sensor area to another.

However, like other wide area airborne surveillance systems, the ARGUS collects a huge amount of information, potentially creating data transfer delays. It can store up to one million terabytes of data per day, recording the equivalent of 5,000 hours of high definition video.[136] Such a large amount of data requires a lot of bandwidth to transfer from the aircraft to a ground base where analysts can review it.

One potential method to smooth this process is through an aerial command center, known as Integrated Sensor is Structure (abbreviated to ISIS, at least until that acronym became commonly associated with a terrorist group). Directed by DARPA and the United States Air Force Research Laboratory, the project aims to develop a high-altitude airship that would carry sensors, including a radar with a range of 600 kilometers, and could link to ISR drones.[137] While the ARGUS films from a maximum height of 20,000 feet,[138] the Integrated Sensor blimp will fly over 60,000 feet, out of the range of most anti-aircraft weapons. Additionally, from that height, it could track aerial objects along with those on the ground. Held aloft by helium, and powered, at least in part, by solar energy, the airship could remain aloft for long stretches of time, perhaps multiple years. In April 2009, DARPA awarded a $400 million contract to Lockheed Martin and Raytheon to produce a prototype.[139]

A functional Integrated Sensor blimp could address some of the technical problems of Gorgon Stare. Like the ARGUS, the Integrated Sensor system uses a single radar and sensor array covering a wide area, so it would not lose track of an object as it moved from one sensor area to another, though its high-altitude position reduces the ability to provide detailed visual imagery to observers. As a blimp flying above 60,000 feet, an Integrated Sensor is Structure system could communicate more easily with a satellite, improving upon ARGUS or Gorgon Stare's difficulties with data transmission. Additionally, the airship could function as an informational mothership, gathering data from other ISR platforms in the area, and relaying that information to analysts on the ground. By combining data from the sensors of Gorgon Stare, the high definition video of ARGUS, and the powerful radar of the Integrated Sensor blimp in one ISR command center, battlefield commanders and intelligence analysts could gather a detailed portrait of a designated area. However, DARPA's Integrated Sensor is Structure efforts, and similar blimp projects, such as the United States Army's Long Endurance Multi-Intelligence Vehicle (LEMV) and the Air Force's Blue Devil 2, have yet to produce a workable system.[140]

A functional high-altitude blimp would help address the technical problem of delayed data transmission, but not the conceptual problem of the bottleneck created by data analysis. A fully operational Gorgon Stare, ARGUS, or alternative wide area airborne surveillance system could produce visual, infrared, and electro-optical information about a 100 square kilometer area, covering an entire town or a significant portion of a large city. (For reference, Baghdad covers 734 square kilometers.) These systems produce an immense amount of data, requiring dozens of human observers to monitor a sparsely populated area, and hundreds to watch and analyze an urban center bustling with activity. Air Force officials working on the ARGUS-IS project have reached out to sports broadcasters and reality show producers seeking advice on how to monitor many simultaneous video feeds.[141]

The United States already has difficulty keeping up with the demand for airborne surveillance, especially in active theaters like Afghanistan. ISR sorties undertaken by manned and unmanned aircraft quintupled over Afghanistan from approximately 500 per month in the first quarter of 2009, to over 1,500 per month in mid-2010,[142] to over 2,500 per month in the first nine months of 2011.[143] In the more limited campaign against ISIS in Iraq and Syria, the United States flew 1,700 ISR sorties between August and December 2014 alone.[144] Analyzing the data from each of these flights requires considerable time and effort, even when officers preselect a target for surveillance rather than instructing the drone to survey a larger area.

A wide area airborne surveillance system could reduce the number of flights necessary to gather the same amount of information, but the true advantage of Gorgon Stare or ARGUS is the ability to observe many targets at once, including areas users do not know are important in advance. However, accomplishing this would require observers to actively monitor every piece of information the system acquires. Highlighting this problem at a conference in November 2010, General James E. Cartwright, the Vice Chairman of the Joint Chiefs of Staff, lamented "an analyst sits there and stares at Death TV for hours on end, trying to find the single target or see something move. It's just a waste of manpower."[145]

This demonstrates that current technologies cannot yet overcome states' informational and responsibility disadvantages in asymmetric warfare. Terrorists and insurgents have a plethora of potential targets. Attacking civilians or local government officials in any populated area disrupts normalcy, while attacking government forces or anyone working with them in any location imposes a cost on the counterinsurgents. Counterinsurgents have to protect all of these targets at once, without knowing where insurgents will attack, or, in many cases, who is an insurgent and who is a civilian. Without good human intelligence to direct analysts' focus, ISR systems just gather a flood of information that may or may not be important. Given the limited ability of operators to monitor all of this data at once, they might not learn a particular piece of information is relevant until after an attack.

Therefore, it would be extremely valuable if advancements in information gathering were accompanied by advancements in information processing. Instead of human analysts staring at live video feeds in case something might happen, computer software could monitor many feeds simultaneously, and alert human analysts if anything requires their attention. This would allow a few people to monitor a large area, and no humans would waste time watching locations where nothing is moving. As discussed in Chapter 5, this could facilitate an information-focused strategy of robotic warfare, in which comprehensive situational awareness facilitates faster, more accurate decision making.

As of yet, software capable of autonomously monitoring a large area does not exist, though there have been considerable advancements in the field of computer vision, which indicates such a software package is possible. However, computer vision already plays a significant role in ISR systems. Object recognition and motion tracking allow systems like the ARGUS to keep a visual window focused on a person or vehicle as it moves through the larger image. This requires distinguishing the object from a constantly changing background and from similarly shaped objects that enter the frame. Nevertheless, while the system can autonomously follow a designated item, a human operator must first select a target for the system to track.

Developing software capable of watching "Death TV for hours on end" in support of, or in place of human analysts could help security services anticipate and stop attacks, and track down attackers more easily if they cannot stop them in advance. To prevent disruption, counterinsurgents need to protect many potential targets at once. In larger areas, this becomes cost prohibitive, since it is almost impossible to know when and where insurgents will attack. Oil pipelines, power lines, railroads, and other infrastructure are thus attractive targets for insurgents, because it is difficult to defend all of them at once, and a breach anywhere along the route can disrupt the flow of resources, electricity, or goods. However, a computer, monitoring video and infrared sensors, could effectively watch an entire length of pipeline or railroad, alerting human operators to suspicious activity, such as people approaching a remote area at night. If security forces cannot arrive in time to prevent the attack, the computer could track any people leaving the area following an explosion, allowing soldiers to intercept them before they can blend back in to the civilian population.

Attacking government forces as they move along roads is among the most successful insurgent tactics. Convoys of troops or supplies are pretty conspicuous. It is therefore easier for insurgents to know the route of a convoy than for counterinsurgents to know the location of ambushes or explosives hidden next to, or buried under, a road. A wide area airborne surveillance system could monitor roads in front of convoys, spotting people that may not be visible to soldiers traveling along the road, and use object recognition to autonomously search for weapons. Additionally, if computer vision software can learn to recognize basic actions in addition to objects, cameras could monitor every stretch of road in a given area, alerting human operators when they spot someone potentially planting explosives. The action in question may be innocuous, and the system might not be able to recognize every type of explosive device, but it could alert a human analyst to the suspicious activity, who could then zoom in closely and review the video.

Meanwhile, data gathered by airborne ISR systems, including Gorgon Stare and ARGUS, could inform after-the-fact analysis. With wide area airborne surveillance systems and Integrated Sensor blimps capturing images of everything within an area of interest, there would be a record of every IED explosion, every attempted ambush, and every outdoor movement of people and vehicles. Analysts could closely analyze a single event to identify mistakes and develop countermeasures, or analyze multiple events of the same type, looking for patterns.

For example, after an IED explodes under a patrolling vehicle, or fails to cause any damage thanks to an Explosive Ordinance Disposal team, analysts could review video from the bomb's location. They could run the video back to the point when an insurgent placed the IED, and then follow him to his next location or reverse the video further to discover his previous location, thereby potentially finding where the bomb was made. Real-time analysis would be more valuable, by enabling ground forces to arrive quickly when an insurgent plants the bomb, arrest the bomber, and dispose of the ordinance before it can harm anyone. Nevertheless, the opportunity to discover insurgent hideouts, bomb-making factories, and weapons caches demonstrates the vast potential of wide area airborne surveillance systems, even while real-time data analysis remains a bottleneck.

SMALLER INFORMATION-GATHERING ROBOTS

In addition to ISR sensor systems mounted on Global Hawks, Sentinels, Predators, and other large UAVs, advanced militaries utilize smaller robots to gather information. On the ground and in the air, these robots operate on the tactical level, improving soldiers' battlefield awareness. This can reduce counterinsurgents' informational disadvantage, especially in urban environments, as soldiers move along roads, through alleyways, and into buildings.

The RQ-11 Raven, made by AeroVironment, is a small unmanned plane that carries video, electro-optical, and infrared cameras, enabling it to gather information day and night. Weighing between 4.2 and 4.8 pounds with a wingspan of 4.5 feet,[146] it uses an electric motor to fly up to a maximum of 15,000 feet above sea level, though it more frequently operates and achieves maximum performance around 500 feet above the ground, and can remain aloft for up to 90 minutes. With a flying speed between 28 and 60 mph, the Raven's range is effectively 10 kilometers. The three cameras transmit information to a ground control station, which can display the images in real-time or store them for future analysis. Together, a Raven and its ground control station cost approximately $250,000. The United States granted AeroVironment a contract to produce 2,358 Raven systems, and additional units have been purchased by American allies, including Australia, Italy, Denmark, the Netherlands, the UK, and Spain.[147]

Unlike larger UAVs, such as the Predator or Global Hawk, the Raven is carried by and operated by soldiers in the field. To launch a Raven, a person throws it into the air, using an over-the-shoulder motion similar to throwing a javelin.[148] Once in the air, soldiers can direct it manually via the ground control station, or it can use GPS to fly autonomously according to pre-programmed specifications. The Raven also lands autonomously and does not require a prepared landing strip, making it well suited for forward-deployed units, especially in harsh terrain.[149] However, some soldiers have complained the Raven is difficult to launch and crashes often, requiring frequent repairs or replacement.[150]

AeroVironment also makes a smaller "micro air vehicle" known as the Wasp. Less than a foot long, with a wingspan of 28.5 inches, the Wasp weighs only one pound, making it highly portable and easy to throw. It carries two cameras, each approximately the size of a peanut, that can gather information day and night, and, like the Raven, transmits the information it gathers to a ground-based control station.[151] Using an electric motor with rechargeable lithium ion batteries, it can travel at speeds ranging from 20 to 40 mph, reach heights of 1,000 feet above ground level, and fly by remote control or autonomously using GPS and an internal navigation system.[152] Unlike the Raven, the Wasp comes in an "all environment" version capable of full functionality at sea, as well as on land.[153] Each system—plane and control station—costs approximately $50,000.[154]

Small UAVS, like the Raven and Wasp, provide soldiers with the ability to gather information about their surroundings at their discretion, instead of requesting assistance from a nearby Predator or Sentinel and having to wait for a response. With a Raven or Wasp, ground units can look over hills or onto rooftops, scout ahead to the next city block, around a curve on a mountain path, or a few miles down a road, and observe their vicinity from a better vantage point. This gives them lead time to prepare for potential encounters with civilians or enemy fighters, and provides a layout of the terrain in which the interaction could take place. Perhaps most importantly, aerial observation could give soldiers or Marines advance notice of an ambush, or at least help them locate the source of incoming fire and determine the easiest way to counter it.

Tiny helicopter UAVs provide a smaller, more maneuverable alternative to these small planes. Beginning in 2012, British forces in Afghanistan began utilizing a "nano helicopter" drone known as the Black Hornet, produced by the Norwegian company Prox Dynamics.[155] Officially called the PD-100 PRS (for "Personal Reconnaissance System"), the Black Hornet is four inches long and weighs only 16 grams (about half an ounce), easily fitting in the palm of an adult's hand. Despite its small size, it is capable of operating in windy conditions, can fly up to 22 mph, and remain in the air for a maximum of 25 minutes before its batteries require recharging.[156] It launches from a small base station, which, together with the drone, weighs less than a kilogram and can fit inside a pants pocket. Like the Raven and Wasp, the Black Hornet can be piloted remotely using a hand-held controller, follow a pre-programmed course, or utilize GPS to autonomously survey a designated area.[157]

Essentially a flying camera that provides real-time video or still photos with a maximum visual range of 1,000 meters, the Black Hornet provides ISR capability to individual soldiers. While large UAVs, like the Global Hawk or Predator, typically serve theaters and are operated from remote command centers, and smaller UAVS, such as the Raven, typically serve a platoon, the Black Hornet can serve a single squad. Each nano helicopter launches itself, requires minimal training and no pilot experience to fly, and transmits information back to a small display unit.[158] It is thus possible for multiple soldiers in a small, 8–12 person unit to each operate a Black Hornet, looking in multiple directions at the same time, or maintaining ISR capabilities even if they split into smaller sub-units.

Nano UAVs can therefore help small groups of soldiers overcome informational disadvantages as they move over open ground, patrol streets, or raid buildings. These actions are among the most dangerous for counterinsurgents, as they expose soldiers operating in the open to fire from hidden locations. However, by flying ahead of soldiers, nano helicopters can help determine the location of enemy positions, and give advance notice whether people are armed fighters or civilians. According to British Sergeant Christopher Petherbridge of the Brigade Reconnaissance Force in Afghanistan, the "Black Hornet is definitely adding value, especially considering the lightweight nature of it. We use it to look for insurgent firing points and check out exposed areas of the ground before crossing, which is a real asset."[159] Additionally, because they are so small, they can operate inside buildings as well as outside, and are quiet enough to attract minimal attention.[160] In Afghanistan, British soldiers have used them to see around corners and into rooms.[161] The British Ministry of

Defense granted a contract to Prox Dynamics for £20 million to provide 160 Black Hornet systems,[162] and, based on the positive early reviews, other countries will likely follow suit.

Besides aiding soldiers by looking beyond their line of sight, small UAVs can conduct electronic surveillance. In 2011, at the Black Hat and DEFCON security conferences—which feature hackers and computer security professionals—security consultants Mike Tassey and Richard Perkins presented a homemade drone that can spy on both wireless computer networks and cell phones, which they named the Wireless Aerial Surveillance Platform (or WASP—no relation to the small ISR drone by AeroVironment known as a Wasp).[163] This UAV is 76 inches long, with a wingspan of 67 inches, and can remain in the air for 30–45 minutes with a maximum altitude of 22,000 feet.[164] The WASP can hack password encrypted Wi-Fi computer networks, and act as a GSM antenna (Global System for Mobile), which allows it to intercept cell phone calls and text messages.[165] Any cell phone closer to the WASP than a cell tower will connect with the drone first, allowing it to gather any information sent to or from nearby mobile devices.

Tassey and Perkins, who have experience working for the US intelligence and defense communities, built the drone to prompt new developments in electronic security by demonstrating the risks to electronic communications, but it offers apparent intelligence-gathering capabilities as well. A WASP could intercept insurgent communications and gather information off militants' computers without their knowledge. This would allow intelligence analysts to monitor communications insurgents believe to be secret, perhaps as they discuss strategy, identify members, or plan attacks. If a terrorist network discovered the WASP's capabilities, it would still grant counterterrorists an advantage, because avoiding wireless networking or cell phones would hinder the terrorists' ability to communicate.

Not all small information-gathering robots are airborne. The Scout XT, from Recon Robotics, is a "throwbot," a small ground-based robot soldiers can throw over walls or into buildings. It looks like a rolling dumbbell with antennae: a cylindrical tube with a wheel on each end in place of the weights. The Recon Scout weighs 1.2 lbs and can be thrown 100 feet or more. Upon landing, its camera and microphone switch on and transmit data back to a hand-held control unit, which a soldier uses to direct the robot. The Scout includes infrared, as well as ambient light cameras, enabling operation in both dark and light environments.[166] In demonstration videos, the Scout proved its durability by falling 30 feet onto a concrete surface, bouncing, and then rolling along as normal.[167] Together, the robot and its control unit weigh three pounds, making it easily transportable by individual soldiers. In early 2012, the United States Army awarded a $13.9 million contract to Recon Robotics for 1,100 Scouts, the largest order in the company's history.[168]

The tactical advantages of the Scout are similar to those of the Black Hornet, with a few noticeable differences. The Recon Scout is quiet, operating at just 22 decibels.[169] To put that in perspective, a typical refrigerator hums at 40 decibels, and a human whisper is around 30.[170] Therefore, the Scout can look around corners and enter rooms ahead of soldiers without attracting much notice, sending back information that can alert them to potential dangers. Comparatively, the main disadvantage of a Scout is that it rolls rather than flies, which means it cannot climb stairs or view a scene from above. But it is a lot cheaper. Based on recent orders, each Scout costs less than $13,000, while each Black Hornet costs just under $200,000. As a result, Scouts may be more cost effective, especially for organizations with lower budgets than the military, such as law enforcement and first responders.

FIGURE 2.9   An American soldier, preparing to launch a Raven.

GUNFIRE DETECTION

To illustrate how new technologies can overcome informational disadvantages, consider the microcosm of combat between a squad and a sniper. With more people and more guns—and possibly mortars, RPGs, and the ability to request assistance from artillery and aircraft—the squad of soldiers enjoys a large resource advantage over an individual sniper, or two-person sniper team. However, the sniper can threaten the squad because of an informational advantage.

The sniper is hidden, his location unknown to the squad. By contrast, the soldiers are on patrol or advancing in the open. As such, the sniper can capitalize on surprise, shoot at the squad, and continue shooting with relative security until the soldiers discover his location. In settings with considerable cover, such as cities or jungles, the sniper can shoot and quickly move to a new spot. A squad of soldiers from a powerful military could easily defeat a sniper team, if they knew its location. However, as long as the snipers remain hidden, their informational advantage neutralizes the squad's resource advantage.

To overcome this informational disadvantage, the squad could use recently developed gunfire detection systems. These anti-sniper systems utilize sound detection to determine the location of a gunshot. For example, the Boomerang Mobile Acoustic Shooter Detection System (MASDS) from BBN Technologies identifies the location of a shooter to plus-or-minus 15 degrees accuracy within one second of the shot. According to BBN, the Boomerang can detect fire from AK-47s and other small arms at ranges of 50 to 150 meters, and can operate on a vehicle moving up to 60 miles per hour.[171]

An alternative that does not require a separate system is the Robot Enhanced Detection Outpost with Lasers (REDOWL) addition to the commonly used PackBot from iRobot.[172] In conjunction with the Photonics Center at Boston University, iRobot developed a system that combines optic and acoustic sensors to pinpoint the origin of gunfire. It utilizes an algorithm based on human hearing to process acoustic information, as well as daylight and low-light cameras, thermal imaging, a laser range finder, and GPS positioning to locate the shooter, day or night, and shine a laser pointer at the shot's point of origin. In firing range field tests for the US Army's Rapid Equipping Force, the REDOWL system demonstrated a 94% success rate locating the origin of shots from M-16 and AK-47 rifles at more than 100 meters.[173]

While these and other gunfire detection systems have proven their capabilities in controlled demonstrations, the United States has not deployed them to combat zones. The systems are unable to distinguish friendly weapons and calibers from hostile fire, and the robot becomes useless, or potentially dangerous, in a firefight. As shots ring out from all sides, the robot's head goes "into a laser-aiming seizure,"[174] swinging around wildly. This negates its capability and creates the risk it will hit, or shine its laser pointer into the eyes of, a friendly soldier.

Despite these technical issues, development of gunfire detection systems will continue. The ability to locate the origin of gunfire would neutralize the informational advantage that allows a sniper to threaten a squad of soldiers, and would allow soldiers to respond more effectively to ambushes. If the REDOWL or similar systems can remain focused on the first shot, or can learn to distinguish nearby friendly fire from distant enemy fire, they will help powerful militaries assert their resource advantage against hidden foes.

Terrorists and insurgents exploit uneven levels of resolve, responsibility, and information inherent in asymmetric warfare to combat stronger opponents, but robotics can help states overcome these disadvantages. This is already apparent from the American-led asymmetric conflicts in Iraq and Afghanistan, where unmanned systems in the air and on the ground have decreased military casualties, improved the efficiency of targeting, reduced collateral damage, and revolutionized information acquisition. Robotics technology will continue advancing, impacting asymmetric warfare in various ways, though few will capture the public imagination like drone strikes.

CHAPTER 3

Drone Strikes

"Could the use of flying death robots be hurting America's worldwide reputation?"[1] That question comes from the Onion, a satirical website, and while the phrasing is deliberately ridiculous, the underlying idea has merit. Popular sentiment plays a central role in asymmetric warfare, and the United States and other drone-armed governments must factor in public reaction when measuring the costs and benefits of military activity.

Firing missiles from unmanned aircraft can facilitate a relatively inexpensive counterterrorism strategy. Drone strikes often kill terrorist leaders and operatives at a lower cost and with less risk to personnel than alternative methods, potentially reducing their ability to plan and execute attacks. However, these strikes create blowback, as anger over targeted killings, extrajudicial attacks outside of warzones, and collateral or accidental damage to civilians aids terrorist propaganda and recruitment. Any strategically useful drone campaign must balance these two countervailing factors.

THE AMERICAN DRONE CAMPAIGN

In the twenty-first century, the United States has fired missiles from Predator and Reaper drones to strike targets linked to Al Qaeda, ISIS and other jihadist groups, as well as Afghan insurgent networks. The United States does not officially acknowledge these attacks because many are extrajudicial, taking place outside of official warzones such as Afghanistan and Iraq, and lacking a clear international legal architecture.[2] According to the Drone Wars project at the New America Foundation (NAF), from November 3, 2002—the date of the first strike in Yemen—through the end of 2016, American drones launched approximately 403 attacks in Pakistan, 183 in Yemen, and 41 in Somalia.[3]

The totals are approximate because, given the lack of official statistics, reports occasionally conflict regarding whether the strike came from a UAV or a manned aircraft, or whether American or local government forces were responsible. For example, according to diplomatic cables revealed by WikiLeaks, Yemeni officials claimed responsibility for American airstrikes to avoid a public outcry over the government granting foreign forces permission to launch attacks against Yemeni citizens on their territory, with Yemeni President Ali Abdullah Saleh telling American General David Petraeus "we'll continue saying the bombs are ours, not yours."[4] The NAF's figures thus represent confirmed American drone strikes, which could be considered a low-end estimate. Alternative sources, especially those critical of America's drone campaign, such as the Bureau of Investigative Journalism's "Covert Drone War" project, compile data on any possible drone strike, estimating 424 attacks in Pakistan and up to 249 in Yemen, which effectively provide high-end estimates.[5]

Of the three locations, Pakistan features the best international press coverage and has received the most scholarly attention. The New America Foundation's data "draws only on accounts from reliable media organizations with deep reporting capabilities in Pakistan, including the *New York Times*, *Washington Post*, and *Wall Street Journal*, accounts by major news services and networks (the Associated Press, Reuters, Agence France-Presse, CNN, and the BBC) and reports in the leading English-language newspapers in Pakistan (the *Daily Times*, *Dawn*, the *Express Tribune*, and the *News*), as well as those from Geo TV, the largest independent Pakistani television network."[6] Using these sources, they compile drone attacks in Pakistan through the end of 2016 as follows.

The United States first employed attack drones in Pakistan in 2004, but the UAV campaign began in earnest in 2008. The number of strikes escalated from 36 in 2008, to 54 in 2009, and peaked at 122 in 2010, declining annually thereafter. The initial rise indicates an increasing reliance on UAV strikes to target various insurgent networks operating along the Afghanistan–Pakistan border, along with remaining members of Al Qaeda. The early years saw many civilian casualties, but the percentage of civilian deaths decreased

TABLE 3.1   US Drone Strikes: Pakistan

Year	Drone Strikes	Estimated Total Deaths		Estimated Militant Deaths		Percentage Civilian Deaths[7]		Est. Militant Leader Deaths
		Low	High	Low	High	Low	High	
2004–7	10	154	201	43	76	72.08	62.19	3
2008	36	218	348	157	263	27.98	24.43	11
2009	54	358	708	240	516	32.96	27.12	10
2010	122	525	831	484	777	7.81	6.50	13
2011	72	393	623	317	528	19.34	15.25	5
2012	48	229	365	210	330	8.30	9.59	8
2013	26	120	161	117	156	2.50	3.11	12
2014	22	128	157	128	157	0.00	0.00	5
2015	10	50	61	48	59	4.00	3.28	1
2016	3	8	11	8	11	0.00	0.00	2
Total	403	2183	3466	1752	2873	19.74	17.11	70

TABLE 3.2 US Drone Strikes: Yemen

Year	Drone Strikes	Estimated Total Deaths		Estimated Militant Deaths		Percentage Civilian Deaths[8]		Est. Militant Leader Deaths
		Low	High	Low	High	Low	High	
2002	1	6	6	6	6	0.00	0.00	2
2009	2	85	105	44	64	48.24	39.05	2
2010	1	6	8	2	2	66.67	75.00	1
2011	12	96	127	83	99	13.54	22.05	9
2012	56	410	540	378	504	7.80	6.67	22
2013	25	112	140	91	115	18.75	17.86	10
2014	19	106	163	99	150	6.60	7.98	0
2015	24	91	91	90	90	1.10	1.10	3
2016	43	183	203	180	200	1.39	1.48	7
Total	183	1095	1383	973	1230	11.14	11.06	56

over time, likely due to better intelligence, more selective targeting, and improving skill.

The declining number of UAV strikes and associated deaths after 2010 demonstrates the effectiveness of the campaign, as fewer targets become available. While this could be evidence that counterterrorist measures eliminated a significant percentage of fighters, the decline could also indicate insurgents adjusted their behavior to reduce their vulnerability to aerial attacks. In 2010, leaders of the Pakistani Taliban admitted fear of drone strikes drove their group underground.[9] This suggests the drone campaign in Pakistan has disrupted insurgent operations in the Afghanistan–Pakistan theater by killing fighters and leaders, and denying the remaining members the ability to operate openly.

Meanwhile, as the previous table shows, the drone campaign in Yemen peaked in 2012, with more UAV attacks than in all previous years combined, and spiked again in 2016. While Yemen was the site of the first extrajudicial UAV attack—the Predator-launched strike that killed Ali Qaed Senyan al Harthi in 2002—the American drone campaign began focusing on Yemen after Al Qaeda in the Arabian Peninsula (AQAP) claimed responsibility for the attempted "underwear bombing" on December 25, 2009, in which Umar Farouk Abdulmutallab tried to detonate a plastic explosive called PETN sown into his underpants while on board a Northwest Airlines flight from Amsterdam to Detroit. Beginning with an attack on a suspected AQAP training camp in December 2009, the United States launched approximately 27 drone strikes through June 2012,[10] and at least 27 more in the following seven months, with six attacks in January 2013 alone.[11]

By the end 2016, the 183 strikes killed an estimated 1,095–1,383 people in Yemen, 973–1,230 of whom the New America Foundation identified as militants, for a civilian casualty rate of 11.06%–11.14%. The Bureau of Investigative Journalism (BIJ) estimates a civilian casualty rate in Yemen of 11.93%–12.74% for 124–144 "confirmed drone strikes," and a similar rate of 10.16%–12.50% when including "possible extra drone strikes" for a total of 212–249.[12] These are somewhat lower than NAF's estimated 17.11%–19.74% and BIJ's estimated 16.97%–24.14% civilian casualty rate in Pakistan.

In July 2016, the United States government publicly acknowledged the drone campaign for the first time, releasing internal statistics. The Obama administration claimed a civilian casualty rate of

2.63%–4.30% for 473 extrajudicial strikes (the government's data did not specify location) from 2009 to 2015.[13] The large discrepancy with non-governmental estimates likely reflects the government's incentive to downplay harm to civilians.

KILL LISTS

The strikes in Yemen have killed an estimated 56 militant leaders—primarily senior members of Al Qaeda in the Arabian Peninsula, such as the organization's head of media, Ibrahim al Bana, on October 14, 2011[14]—while the attacks in Pakistan have killed about 70. These individuals were likely among those on secret kill/capture lists the United States began developing in the aftermath of September 11th.

On September 14, 2001, Congress passed an Authorization for Use of Military Force by a nearly unanimous vote—Representative Barbara Lee, a Democrat from California, was the only person in either house who voted Nay[15]—declaring "the President is authorized to use all necessary and appropriate force against those nations, organizations, or persons he determines planned, authorized, committed, or aided the terrorist attacks that occurred on September 11, 2001, or harbored such organizations or persons, in order to prevent any future acts of international terrorism against the United States by such nations, organizations or persons."[16] With this authority, President George W. Bush issued a presidential finding that directed the CIA to hunt down and kill or capture terrorists without requiring presidential approval for each operation.

The Bush administration created a list of about two dozen names, featuring Osama bin Laden, Ayman al Zawahiri and other senior Al Qaeda figures, along with leaders of affiliated organizations, known as the "high value target list." It authorized the CIA to kill individuals on the list if capture was impractical and civilian casualties could be minimized; though "impractical" and "minimized" were mostly left up to interpretation on a case-by-case basis. The Bush administration officially declared individuals on the list to be "enemy combatants," placing them outside the Geneva Conventions, which only cover civilians and members of participating state militaries. Similarly, enemy combatants are not covered by domestic and international prohibitions on assassination.[17] Though it started with about 25 names, the original high value target list was never supposed to be a complete register. Both the White House and CIA could add to it.

In 2010, the Obama administration directed the National Counterterrorism Center (NCTC) to consolidate the CIA's high value target list, which had expanded throughout the 2000s, and a parallel list maintained by the Defense Department's Joint Special Operations Command (JSOC), into a constantly updating database called the "disposition matrix." The disposition matrix catalogues information about terrorism suspects, including personal history, location, known associates and affiliated organizations, as well as strategies for dealing with each suspect in a variety of situations. For example, as a former counterterrorism official told the *Washington Post*, "if he's in Saudi Arabia, pick up with the Saudis. If traveling overseas to al-Shabaab [in Somalia] we can pick him up by ship. If in Yemen, kill or have the Yemenis pick him up."[18]

This database is less ad hoc than the original high value target list, which both improves the process and demonstrates that targeted killing of individual terrorist suspects will be a central feature of American counterterrorism strategy for the indefinite future. The disposition matrix is more expansive, including more information about a greater variety of suspects, many of whom are not leaders, skilled bomb-makers, and others typically considered high value. Nevertheless, under the Obama administration, the United States employed a top secret "nominations" process directed by the White House that designated individual terrorist suspects for kill or capture. Capture is rare. More often, when the United States locates someone on the kill/capture list, the response is a drone strike.[19]

On September 7, 2015, British Prime Minister David Cameron revealed the UK has its own kill list. On August 21, 2015, the UK executed its first successful drone strike outside of a warzone, killing ISIS fighter Reyaad Khan in Syria with a missile fired from a Reaper. The Royal Air Force had previously used drones to strike targets in Afghanistan, but only when British or allied forces were threatened by fighting on the ground.[20] Khan, along with Mohammed Emwazi—an ISIS member nicknamed "Jihadi John" who appeared in multiple internet videos speaking English as he beheaded prisoners—was reportedly on a list of at least five individuals created by the UK National Security Council.[21] Emwazi, who died in an American drone strike in November 2015,[22] was a British citizen, as were Khan and Ruhul Amin, a suspected ISIS fighter who was killed in the strike targeting Khan.[23]

The drone strike and revelation of the kill list ignited a debate in the British Parliament, with critics objecting to an expansion of British military activity into Syria and the targeting of British citizens, both without direct Parliamentary authority. In response, Prime Minister Cameron argued Khan was actively planning terrorist attacks in the UK and other Western countries, which meant the drone strike was justified by "the UK's inherent right to self-defense."[24] Additionally, Defense Secretary Michael Fallon asserted the strike, which was approved by Britain's Attorney General Jeremy Wright, "involved hours of surveillance and a great deal of planning to comply with the rules of engagement that we set that there should be no civilian casualties or other damage."[25]

The attack that killed Khan and Amin indicates the American drone campaign has set a precedent other countries will follow. The UK is now the third country, after the United States and Israel, to conduct targeted drone strikes outside of warzones against individuals the government identifies as terrorist suspects, but it will not be the last. Most, if not all governments can see the appeal of attacking suspected enemy combatants without risking personnel. As a technologically advanced country and close American ally, Britain was a logical candidate to adopt the tactic of extrajudicial drone strikes. Its rules of engagement are similar, though the UK currently specifies zero tolerance for civilian casualties, while the United States calls for minimizing risk to civilians. However, other countries, such as Russia or Iran, might apply different standards.

AMERICAN DRONES KILLING AMERICANS

On September 30, 2011, an American drone strike killed Anwar al Awlaki in Yemen. In addition to assisting with operational planning, al Awlaki was the public face of AQAP, broadcasting the group's message in sermons on the internet, and directly communicating with underwear bomber Umar Farouk Abdulmutallab, Fort Hood shooter Nidal Malik Hasan, and others.[26]

Killing Awlaki created some controversy because he was an American citizen. Though he was not on American soil and held dual American–Yemeni citizenship, Awlaki was arguably entitled to the same Constitutional rights as any other American citizen, including the Fifth Amendment's guarantee of due process, and the right to trial by jury specified in the Sixth Amendment. The Obama administration countered it had the legal authority to classify Awlaki as an enemy combatant under the 2001 Authorization of Military Force because he was taking part in the war between the United States and Al Qaeda, he posed a significant threat to Americans, the Yemeni government was unwilling or unable to stop him, and it was not feasible to capture him. However, the full legal memo laying out this case—which also cites precedent of American courts allowing the government to detain and prosecute American citizens who had joined enemy forces in military courts as if they were non-citizen enemies—remains secret.[27]

Critics assert killing Awlaki was an abuse of power that violates the Constitution and sets a dangerous precedent. Arguably, the Obama administration did not even follow its own standards for drone strikes, which limit attacks to individuals who pose an "imminent threat." Though few doubt Awlaki worked with AQAP, he did not execute any attacks himself, and the United States did not claim Awlaki was in the process of orchestrating an attack at the time of his death. As a result, Hina Shamsi of the ACLU argues the killing of Awlaki represents "one of the most extreme and dangerous forms of authority that the executive branch can claim – the power to kill people based on vague and shifting legal standards, secret evidence and no judicial review after the fact."[28] Additionally, some analysts, such as Middle East expert Juan Cole—who admitted "satisfaction" at the killing of a "notorious proponent of radical terrorism against the United States"—expressed concern with the precedent and potential for abuse, arguing the United States should have tried Awlaki in absentia before sentencing him to death.[29]

Awlaki is not the only American citizen the United States has killed in a drone strike, though he was the only one the United States targeted directly. A strike in Pakistan on November 16, 2011 killed Jude Kenan Mohammad, an American citizen indicted in North Carolina for conspiracy to commit terrorism, though the United States did not know he was there when launching the attack.[30] The same attack that killed Awlaki also happened to kill Samir Khan, who grew up in a middle-class household in Queens, New York. In his later teenage years, Khan participated in online jihadist forums, and tried to use his computer skills to assist various groups online. Khan left the United States for Yemen when he was 23, joined AQAP and founded *Inspire*, Al Qaeda's online magazine. He told ABC news in 2010 "I am proud to be a traitor." The United

States also killed Anwar al Awlaki's 16-year-old son, Abdulrahman, in a separate attack in Yemen two weeks later in what some American officials called a mistake.[31]

A January 2015 attack against an Al Qaeda compound in Pakistan killed Ahmed Farouq, an American citizen who had become the deputy leader of Al Qaeda in the Indian subcontinent. The United States identified the compound as a target, but did not know exactly who was inside at the time of the strike. The same attack also inadvertently killed two hostages, American Warren Weinstein and Italian Giovanni Lo Porto, both aid workers. American officials stated that intelligence indicated, with "near certainty," there was "no reason to believe either hostage was present."[32]

Another attack in Pakistan around the same time killed Adam Gadahn, who went by the *nom de guerre* Azzam the American. Gadahn was born in Oregon, grew up in California, converted to Islam at age 17, and moved to Pakistan in 1998. He worked as a member of Al Qaeda's "media committee," acting as a translator, cultural interpreter and video producer, and appeared in at least five English-language videos released online in which he urges Americans to support the jihadist cause.[33] In 2006, the United States filed treason charges against Gadahn, making him the only American charged with treason since World War II. As with the attack that killed Farouq, the United States fired at an Al Qaeda compound, but did not know in advance Gadahn was inside.[34]

Despite the controversies surrounding the question of the United States killing American citizens abroad, the drone campaign retains the support of the American people. An AP–GfK poll conducted in April 2015 found nearly three-quarters of Americans believe it is acceptable for the United States to kill an American citizen abroad with a drone strike if that person has joined a terrorist organization, indicating support for operations like the one that deliberately killed Awlaki. However, only 47% agreed it was "appropriate to use drones to target terrorists overseas if innocent Americans might be killed in the process." Those opposed included the 13% of the public that oppose all drone strikes, as well as 43% of the respondents who originally supported or expressed neutrality towards drone strikes in general.[35] Though the public split on this question, a plurality still favored drone strikes targeting terrorists that might accidentally kill innocent Americans, like the attack that killed Farouq, Weinstein and Lo Porto.

SIGNATURE STRIKES

In June 2015, the American drone campaign claimed its highest value target, killing Nasir al Wuhayshi, the leader of AQAP, in Yemen.[36] Al Wuhayshi united jihadists fighting in Yemen and Saudi Arabia, bringing them together under the Al Qaeda banner and transforming the group into a local and international force.[37] In addition to publishing *Inspire*, AQAP is responsible for more attempted attacks against Western targets since 2008 than Al Qaeda Central. Along with equipping the underwear bomber, AQAP placed three bombs made with PETN—a difficult-to-detect plastic explosive—hidden inside printer cartridges on cargo planes bound for Western cities in 2010. One caused a fire on a UPS flight shortly after it took off from Dubai towards Cologne that killed two crew members on September 3, 2010.[38] Acting on a tip from Saudi intelligence, authorities found the other two on October 29, on planes bound for Chicago. The bombs appeared designed to detonate in the air as the planes approached their destinations, destroying the aircraft and raining debris down on whatever was below.[39]

The United States killed Al Wuhayshi with a controversial technique known as a "signature strike."[40] Unlike attacks in which American intelligence identifies and deliberately targets a specific individual, signature strikes target people whose location and actions "match a pre-identified 'signature' of behavior that the U.S. links to militant activity."[41] Critics argue that even when signature strikes kill members of a terrorist organization they are morally questionable, because many "individuals killed are not on a kill list, and the government does not know their names."[42] Additionally, some assert that signature strikes cause significant backlash, angering people in targeted countries, primarily Pakistan, which makes them less likely to cooperate with the United States and aids terrorist recruitment.[43] However, while public opinion research shows considerable anger in Pakistan over the American drone campaign,[44] there is no evidence this anger derives from a specific objection to signature strikes, rather than a more straightforward objection to the United States killing people in Pakistan.

Signature strikes are an attempt to overcome the United States' informational disadvantage. Al Qaeda know the identity and location of group members, but the United States does not. Drones can help compensate with aerial intelligence, but while they can track someone identified by more traditional intelligence techniques, it is hard to identify a specific person from high in the air. Given this difficulty, the United States has supplemented efforts to follow and target known individuals by monitoring areas with militant activity and attacking based on observed behavior rather than specific identities.

This leads to some successes, such as the attack that killed al Wuhayshi, but also some notable failures. For example, an August 2012 strike in Yemen killed an imam who had been working to dismantle Al Qaeda in the Arabian Peninsula.[45] Killing innocents is morally problematic, and in this case a significant strategic setback. Muslim religious leaders speaking out against jihadist organizations can counter the groups' ideological appeal.

Additionally, since signature strikes target behavior rather than specific individuals, they often kill low-level militants who are easily replaceable, which is a small benefit relative to the risk of killing civilians or potential allies like the Yemeni imam. Nevertheless, signature strikes' ability to kill high value targets such as al Wuhayshi and Adam Gadahn create enough benefit that the United States will probably continue employing the technique, at least to some degree.

STRATEGIC OPTIONS

When the United States identifies an individual as an active member of a terrorist organization—whether by locating someone on the kill/capture list, or observing a behavioral signature indicating association with a militant group—or discovers someone is plotting to execute a terrorist attack against an American or allied target, there are at least five potential responses:

(1) Drone strike
(2) Airstrike from a manned aircraft
(3) Ground raid to capture or kill the suspect
(4) Encourage local forces to handle the situation
(5) Leave the suspect alone and focus on anti-terrorism

Drone Strikes vs. Attacks from Manned Aircraft
American drone strikes sometimes kill civilians, including 12 people on their way to a wedding in Yemen in December 2013,[46] but this is a problem for any weapon fired from distance, not drones per se. Manned and unmanned aircraft use similar missiles and targeting equipment, and are therefore equally likely to attack the wrong

target. The United States does not deliberately target civilians as part of asymmetric warfare against terrorist and insurgent groups—there is no strategic value in doing so—but the United States is willing to accept some civilian deaths as collateral damage. In the few instances in which American aircraft fired upon purely civilian targets, the reason is faulty intelligence, and both manned and unmanned planes are equally subject to this problem.

Robotic airplanes do not need to eat, sleep, or use the bathroom. Ground-based drone operators can attend to bodily needs, or change shifts due to fatigue. This means UAVs can remain in the air longer than any manned aircraft, with pilots constantly operating at peak capacity, which allows them to be more selective about the timing of attacks and reduces the likelihood of error. Due to limited flight time and fear of enemy fire, the pilot of a manned plane is more likely to attack a target when the opportunity arises. Unmanned planes, by contrast, can remain in the air for 24 hours or more, allowing them to wait for greater certainty about a target's identity, and for targeted individuals to be isolated from non-combatants.

As a result, drone strikes cause fewer civilian deaths than attacks from manned aircraft. As the following table shows, an increasing reliance on drone strikes coincided with a decrease in civilian casualties in Afghanistan. According to statistics from United States Air Force Central Command, weapons fired from unmanned aircraft increased by 42% from 2011 to 2012, going from 5.45% of total airstrikes to 12.37%. Meanwhile, civilian casualties (including both deaths and injuries) declined by 42%, while civilian deaths from airstrikes declined by 46%. Though 2012 featured fewer total weapons released by aircraft, this cannot explain the decline in civilian casualties, as the rate of both civilian casualties and civilian deaths per airstrike decreased.

These numbers come from Afghanistan, an active military theater, while the tables on the American drone campaign presented above feature missiles fired by unmanned aircraft outside of official warzones.[47] The United States military primarily conducts strikes in warzones, while the CIA executes many of the strikes outside active theaters of war, though in recent years the military has taken on a greater share of the airstrikes, especially in Yemen. The United States has not been firing missiles from manned planes in Pakistan or Somalia—and while American activity in Yemen includes attacks from both manned and unmanned planes, statistics on those

TABLE 3.3 United States Airstrikes in Afghanistan 2010–12

Year	All Weapon Releases from Aircraft[48]	Weapon Releases from UAVs[49]	Percentage of Weapon Releases from UAVs	Civilian Casualties from Air Attacks[50]		Civilian Casualties Per Weapon Release (%)	
				Casualties	Deaths	Casualties	Deaths
2010	5102	279	5.47	306	171	6.00	3.35
2011	5411	294	5.43	353	235	6.52	4.34
2012	4092	506	12.37	204	126	4.99	3.08

manned airstrikes are unavailable—which makes it impossible to compare civilian casualties from extrajudicial drone strikes to those from manned aircraft.

However, the rates of civilian deaths from drone strikes in Pakistan (17.11%–19.74%) or Yemen (11.06%–11.14%) are considerably higher than the United States military's rate in Afghanistan (about 4.22%, based on the above data from 2010–2012). This could be because the military provides a more generous accounting of its own activities relative to the independent assessment of scholars and journalists, or reflect the likelihood American intelligence is more accurate and extensive in locations with a large overt ground presence relative to areas of covert activity. However, comparing one year of military statistics to another from the same source shows a decline in the rates of both civilian deaths and casualties when drone usage spiked, suggesting UAVs are better at minimizing civilian casualties than manned aircraft.

Drone Strikes vs. Ground Raids

While drone strikes are often cheaper than ground raids, they eliminate the possibility of capturing suspects. "Snatch-and-grab" missions risk soldiers' lives, and captured terrorists need to be imprisoned and perhaps put on trial. Both cost time, effort, and money, and risk giving a terrorist a public platform. Additionally, imprisoning terrorist suspects may come with political costs. The United States has faced considerable criticism for holding suspects without trial at Guantanamo Bay and secret CIA "black sites."

However, killing, rather than attempting to capture suspects, forfeits potential intelligence gains. Destroying targets from the air eliminates the chance to interrogate them for information or find material revealing details about the group's membership, finances, strategy, and intended targets. Documents and hard drives captured in the US raid on Osama bin Laden's compound in May 2011 provided a "treasure trove" of information that helped the United States understand and further weaken Al Qaeda.[51]

On October 5, 2013, United States forces captured Abu Anas al Libi in Libya.[52] Al Libi was a senior Al Qaeda operative wanted by the United States for his role in the 1998 embassy bombings in Kenya and Tanzania. He pleaded not guilty to terrorism charges in a federal court in New York City on October 15, 2013, and died in a New York hospital from complications related to liver cancer in January 2015, shortly before the start of his trial.[53] Prosecuting rather than killing al Libi avoided criticism associated with extrajudicial force, and, given his affiliation with Al Qaeda dating back to the 1990s, interrogating him likely yielded valuable information.

However, in the same month the United States captured al Libi, American forces failed to capture a leader of al Shabaab in Somalia. United States Navy SEALs faced unexpected resistance from al Shabaab fighters, and pulled back to reduce the risk of American deaths and widespread civilian casualties.[54] These two October 2013 ground raids demonstrate the tradeoff between undertaking a risky capture mission, which can produce intelligence and lead to legal prosecution, and executing a drone strike, which is more likely to neutralize the target, but eliminates the possibility of intelligence gains.

Ground raids also offer the possibility of greater precision. Ground troops can identify their target in person, and avoid shooting civilians in the target's proximity. The bin Laden raid illustrates the potential advantages of ground raids relative to airstrikes. Fewer civilians died than almost certainly would have if the United States bombed bin Laden's compound, and the Special Operations unit known as SEAL Team 6 was able to acquire valuable intelligence.

However, there is a risk ground operations could fail disastrously. If the bin Laden raid represents a cleanly executed ground raid that highlights the potential benefits, the October 1993 attempt to capture senior members of the Somali National Alliance in Mogadishu illustrates the downsides of failure. In what is popularly known as the Black Hawk Down incident, American Special Operations forces ended up in a prolonged firefight with Somali militants and an unruly mob. As many as 500 Somalis died and many more were injured, most of whom were civilians, and the incident ended with 18 dead American troops dragged through the streets of Mogadishu, images of which appeared on television around the world.[55] Compared to drone strikes, ground raids' potential upside and downside are both greater.

Working with Local Governments

Instead of launching airstrikes or deploying Special Operations forces, the United States could work through local governments, asking them to kill or capture suspected militants. Governments exercising authority over their own territory would create less anger against the United States, and potentially less of a backlash overall. However, the United States already works with the Pakistani government, worked with the government of Yemen until it collapsed in January 2015,[56] and Somalia is a failed state lacking a central government capable of securing its territory.

In both Pakistan and Yemen, the groups America targets with drone strikes are enemies of the local government as well. Al Qaeda, the Pakistani Taliban, the Haqqani network, and other groups based in the Federally Administered Tribal Areas (FATA) near the Afghan border oppose the government of Pakistan. The Pakistani military has launched numerous operations in the FATA, and, from 2003 to 2015, groups based there killed over 6,000 Pakistani soldiers and 20,000 Pakistani civilians.[57] Similarly, AQAP opposed the government of Yemeni President Ali Abdullah Saleh before he was ousted by 2011 Arab Spring protests, and opposed Saleh's successor, Abdu Rabbu Mansour Hadi, before he resigned under pressure from Houthi rebels in 2015. Under both Saleh and Hadi, Yemeni forces launched attacks against AQAP, and the government arrested and jailed Al Qaeda operatives.[58]

Pakistan, and especially Yemen, have less powerful militaries than the United States, which means they cannot attack with the same precision or without accepting greater risk to their soldiers, and both have supported American attacks on their territory. In 2010, a diplomatic cable released by WikiLeaks showed President Saleh giving the United States an "open door" for drone strikes in Yemen, and agreeing to take responsibility for the attacks to avoid the potential political problems that could arise from Yemeni citizens finding out foreign forces were launching airstrikes on their territory.[59] The Hadi government continued cooperating with the United States, but since Hadi's resignation, it is no longer clear America has the permission of the local governing authority to conduct drone strikes, and the United States removed its remaining forces from Yemen in March 2015.[60] The ongoing civil conflict, and lack of a central government, means Yemen more closely resembles a failed state like Somalia, where there is no authority capable of killing or capturing most suspected terrorists. Though the United States pulled back, American drones have continued launching strikes over Yemen, including the missile that killed AQAP leader Nasir al Wuhayshi, most likely taking off from a base in Saudi Arabia.[61]

Secret CIA documents and Pakistani diplomatic memos obtained by the *Washington Post* in 2013 revealed the Pakistani government, like the government of Yemen, endorsed American drone strikes on its territory. Additionally, Pakistani officials received CIA briefings before the United States launched attacks.[62] Some of the earliest Predator flights over Afghanistan and Pakistan took off from Pakistani airstrips, though the United States moved these launches to Afghanistan in an attempt to reduce the appearance of Pakistani complicitness.

Pakistani officials publicly oppose drone strikes, which aligns with Pakistani public opinion.[63] However, Pakistan continues accepting billions in United States aid,[64] the two countries' intelligence agencies continue sharing information,[65] and Pakistan has the ability to shoot down foreign drones on its territory but chooses not to. At the very least, Pakistan tacitly supports the American drone campaign, despite public statements otherwise. Nevertheless, in 2013 the United States agreed to limit attacks on Pakistani soil to high value targets in response to criticism from Pakistani military officials that frequent attacks, including signature strikes, risked creating more enemies than they killed.[66]

When the local government is both willing and able to arrest or kill terrorist suspects on its own, the United States usually prefers not to intervene. For example, in April 2015, Saudi Arabia announced it had arrested 93 people with ties to ISIS and disrupted a plot to attack the American embassy in Riyadh with a suicide car bomb.[67] However, countries have accused United States counterterrorists of arresting suspects on their territory without permission. In 2009, an Italian court tried in absentia and convicted 22 CIA operatives and a United States Air Force colonel on kidnapping charges for the 2003 capture of Hassan Mustafa Osama Nasr on Italian soil.[68] Nevertheless, there is no publicly known instance of the United States firing missiles at terrorist suspects in countries where the local government was willing and able to capture or kill the target themselves and denied the United States permission to operate on their soil.

Focus on Anti-terrorism

It is widely assumed the United States will attack suspected terrorists in some manner, and drone strikes are an attractive option due to the costs and benefits relative to manned airstrikes, ground raids, and working through local governments. However, the United States could choose to focus exclusively on disrupting active plots and strengthening homeland defenses, rather than also killing suspects in an attempt to degrade enemy networks. Refraining from extrajudicial killings would reduce a source of political opposition to the United States.

After the September 11th attacks, the United States improved anti-terrorism defenses. The newly created Department of Homeland Security (DHS), Transportation Security Administration (TSA), and National Counterterrorism Center (NCTC) worked to strengthen security at airports and other ports of entry, as well as various symbolic and soft targets. These defenses are hardly foolproof—for example, in 2015 tests, undercover DHS investigators were able to sneak mock explosives or banned weapons through TSA check points 67 out of 70 times[69]—but their combined efforts, along with new protections set up by local police departments and private security firms, have hardened many potential targets.

To protect against the threat of terrorism, the United States increased intelligence capacity and improved intelligence sharing, both among US agencies and with foreign partners. Post-9/11 reforms aimed to break down information silos and increase cooperation between the FBI, CIA, and Department of Defense, and created over 100 Joint Terrorism Task Forces (JTTFs) to coordinate activity among federal, state, and local intelligence and law-enforcement agencies.[70] The United States also increased intelligence sharing with international partners such as the UK, Germany, Israel and Saudi Arabia. In 2015, America spent approximately twice as much on intelligence as it did in 2001.[71]

In the aftermath of 9/11, the government expanded intelligence agencies' powers. In particular, the National Security Agency (NSA) increased collection of telephone metadata (phone number called and the length of the call, but not the content) and monitoring of internet activity. In a program known as PRISM that began in 2007, the NSA—along with its British equivalent, GCHQ—gathers information on internet users with assistance from nine major technology companies, including Google, Facebook, Microsoft, Yahoo, and Apple. Information from these secret programs, as revealed by former NSA contractor Edward Snowden, contributes significantly to intelligence reports. Data collected by PRISM appeared in the President's Daily Brief 1,477 times in 2012, and provided the lead item in one out of seven NSA intelligence reports.[72] American officials testified in 2013 that the surveillance programs foiled over 50 terrorist plots in the United States and abroad, including a plan to bomb the New York Stock Exchange.[73]

This claim cannot be independently verified, because the details of these alleged plots remain classified, and it is impossible to know whether the attacks would have succeeded if the NSA did not collect information on telephone and internet communications. Additionally, as critics argue, whatever security these surveillance programs provide may not be worth the sacrifice of privacy they require. Nevertheless, when taking into account all post-9/11 reforms and expanded powers, it is safe to say the American intelligence community improved its ability to anticipate and disrupt terrorist attacks before they can succeed.

September 11th also changed the American public's mindset, making the country less vulnerable. Before 9/11, most people assumed airplane hijackers wanted to commandeer a flight to Cuba, use the passengers as hostages in exchange for ransom money, or gain attention and leverage hostages to advance political demands, such as the release of prisoners. 9/11 taught the world that hijackers might kill everyone on board, and others as well. This experience increased citizen vigilance. For example, passengers disabled both "shoe bomber" Richard Reid in December 2001, and the underwear bomber in December 2009, as they tried to ignite explosives.[74] With hardened targets, improved intelligence, and a more vigilant populace, the United States has become less vulnerable to terrorism, which arguably reduces the need to kill or capture members of terrorist organizations abroad.

There is little doubt drone strikes create blowback. Mounting evidence indicates considerable anger in Pakistan and Yemen over these attacks,[75] and both Al Qaeda and ISIS have used the American drone campaign as a recruiting tool,[76] though it is impossible to know how many people this convinces who would not be convinced otherwise. To undermine this recruiting pitch and reduce blowback, the United States could refrain from firing at terrorist suspects abroad, trusting anti-terrorism defenses to disrupt active plots.

Some terrorists will try to attack the United States and its allies in the future because an American drone strike killed someone they love, or because they are angry about the United States violating their country's sovereignty or killing civilians. However, their anger is almost certainly a reaction to the United States using force and killing people, rather than a reaction specific to drones. Any forceful action the United States takes would anger some people, especially those directly harmed by it, and the level of anger is probably a factor of whom the United States kills rather than the weapon used to kill them. If the United States targeted terrorist suspects with missiles fired from manned aircraft instead of drones, or killed a similar number of people in Pakistan and Yemen with ground raids, anti-American groups would design recruitment pitches accordingly.

One of Al Qaeda's central grievances against the United States is American support for Saudi Arabia, Egypt, and other governments in Muslim-majority countries, which indicates working with local governments creates blowback as well. Most people opposed to the drone campaign are opposed to the United States killing people extrajudicially, and would not become more pro-American if the United States switched to an alternative method of killing suspected terrorists and insurgents. Nevertheless, limiting drone strikes to confirmed high value targets, and taking extra care to minimize civilian casualties, could reduce blowback and lessen the appeal of Al Qaeda and ISIS recruitment efforts, thereby weakening anti-American groups and making it easier for locals to cooperate with the United States.

THE STRATEGIC VALUE OF THE DRONE CAMPAIGN

While the United States can point to individual successes, like killing Awlaki and al Wuhayshi, it is unclear if the drone campaign is succeeding strategically. Yemen expert Gregory Johnsen reports Al Qaeda in the Arabian Peninsula grew from 200–300 militants in 2008 to more than 1,000 fighters in 2012, with expanded control in southern Yemen. "In parts of Abyan and Shabwa provinces," he stated, "the organization controls towns in which it has established its own police departments and court systems. It is providing water, electricity and services to these towns. In short, AQAP now sees itself as the de facto government in the areas under its control."[77]

The group's success continued in 2015, despite losing four senior figures to drone strikes, including Ibrahim al Rubaish, AQAP's top ideological leader and mufti (interpreter of Islamic law).[78] Now up to 4,000 members,[79] AQAP took advantage of civil strife in Yemen to gain control of Mukalla, the country's fifth largest city, in April 2015, along with much of the surrounding governorate, Hadramawt.[80] As a result, it controlled a port city on Yemen's southern coast, the city's oil terminal, and the nearby Riyan Airport.[81]

Shortly after, in June, an American drone strike killed AQAP's leader Nasir al Wuhayshi, but the head of military operations, Qasim al Raymi, assumed control of the group.[82] Despite losing its long-time chief, AQAP continued advancing, expanding operations to Taiz, Yemen's second largest city and cultural capital, which is in the west, far from the group's strongholds in the east.[83] This demonstrates UAVs' ability to kill wanted individuals, while casting doubt that a strategy based on targeted killings can successfully neutralize militant groups. AQAP's largest setback came in April 2016, when United Arab Emirates ground forces, backed by the United States, forced it to withdraw from Mukalla.[84]

Research on the value of decapitation strikes shows mixed results. Some studies of Israel's efforts to target the leaders of Hezbollah, Hamas, and Islamic Jihad find leader deaths may reduce terrorist organizations' operational capacity by eliminating highly skilled individuals and forcing groups to devote more of their limited resources to leadership protection. However, decapitation strikes also create martyrs, increase Palestinian and Lebanese anger against Israel, and may result in more terrorist attacks, rather than fewer, as groups lash out in revenge for the death of a beloved leader or as retaliation for a perceived escalation.[85] This suggests targeting leaders may play into, rather than undermine, militant groups' advantages, increasing, rather than decreasing their popular support and determination.

However, other studies find no effect at all. An analysis of Israel's targeted killings in response to the Al Aqsa uprising from 2000 to 2004 found decapitation strikes did not significantly increase nor decrease Palestinian violence. While "targeted assassinations may be useful as a political tool to signal a state's determination to punish terrorists and placate an angry public," the authors argue, there is "little evidence that they actually impact the course of an insurgency."[86]

In some cases, decapitation may weaken a group by eliminating a valuable leader, while in other cases it may inadvertently strengthen the group by paving the way for a talented leader's rise. For example, in 1995 Israel killed Fathi Shikaki, the leader of Palestinian Islamic Jihad, disrupting the organization and leading to a decline in PIJ operations for years.[87] However, after Israel killed Hezbollah leader and co-founder Abbas al Musawi in 1992, Hassan Nasrallah became the general secretary of the group.[88] Under Nasrallah's leadership, Hezbollah expanded its military capabilities, developing a powerful rocket arsenal;[89] successfully fought Israel in a 33-day war in July and August 2006;[90] consolidated its control of southern Lebanon, won seats in the Lebanese parliament, and effectively gained veto power in

Lebanon's cabinet.[91] Killing Musawi clearly did not prevent Hezbollah from becoming more powerful, and more threatening to Israel.

Israel's experience with decapitation strikes may not be applicable to the American drone campaign due to factors specific to the Israeli–Palestinian conflict—such as the prolonged experience of occupation, close proximity of the combatants, passionate local and international opinion, etc.—but larger studies of decapitation strikes show mixed results as well. In a statistical analysis of 298 cases of leadership targeting from 1945 to 2004, international relations scholar Jenna Jordan finds decapitation does not increase the likelihood of organizational collapse beyond baseline expectations, and that larger organizations are more likely to withstand decapitation efforts.[92] By contrast, West Point professor Bryan C. Price's study of 207 terrorist groups from 1970 to 2008 finds experiencing decapitation significantly increases terrorist groups' mortality rate, though defeat did not happen immediately, with 70% of groups that lost their leader surviving more than two years after. Additionally, Price finds the likelihood of success does not depend on group size.[93]

In a study of the decline and end of terrorist groups, terrorism scholar Audrey Kurth Cronin identifies four cases in which killing or capturing leaders played a significant role in their group's demise: "Peru's Shining Path (Sendero Luminoso), the Kurdistan Workers' Party (PKK), the Real Irish Republican Army (RIRA), and Japan's Aum Shinrikyo." However, Cronin identifies 14 other terrorist groups that declined or ended for different reasons, including unsuccessful transitions to a new generation, achievement of goals, transition to a political process/negotiations, loss of popular support, and repression.[94] Additionally, of the four in which decapitation worked, other factors also contributed to the groups' demise—with the exception of Aum Shinrikyo, which depended on a cult of personality surrounding its founder and leader, Shoko Asahara, and declined after his arrest— namely a decline in popular support and repression by the local government.

This shows decapitation alone probably cannot defeat most terrorist or insurgent groups, though it can help bring about their demise when joined with other measures, which helps explain why drone strikes have not defeated AQAP. Though the United States killed multiple senior figures, the chaos in Yemen means the group faces little pressure on the ground, especially since the Yemeni government that opposed AQAP collapsed. The Houthi rebels are

Shia, backed by Iran,[95] while AQAP is Salafist, and has made common cause with local Sunni organizations, attacking Shia mosques and Houthi fighters, and refraining from imposing strict Islamic law in areas it controls to cultivate popular support. Saudi Arabia, also Sunni, intervened in Yemen with the stated goal of restoring the ousted Hadi government, and has trained its fire on the Houthis, avoiding attacks against AQAP, the enemy of its enemy.[96] When AQAP faced a sustained operation by the United Arab Emirates in 2016, it lost ground. However, with support from some elements of the local population, no local government opposing them, limited foreign intervention, and a regional power attacking their main competitor, it is unsurprising AQAP has been able to survive American drone strikes.

Though unlikely to destroy terrorist and insurgent groups on their own, decapitation strikes can degrade their capabilities. In a study of 118 decapitation attempts from 1975 to 2003, political scientist Patrick Johnston finds the number of insurgent attacks and the intensity of conflicts are both more likely to decrease after successful decapitation strikes than after failed attempts.[97] Skilled leaders enhance terrorist and insurgent groups' operational effectiveness, and removing them can aid counterinsurgent strategies.

Additionally, any boost to their resolve derives from attacks in general, rather than successful attacks that kill leaders. As Jenna Jordan argues, "going after the leader may strengthen a terrorist group's resolve, result in retaliatory attacks, increase public sympathy for the organization, or produce more lethal attacks."[98] Groups often react negatively when their enemy targets their leaders, whether or not the attacks kill senior members. However, as Johnston's study demonstrates, reduced insurgent capabilities from successful decapitation strikes can overshadow the associated increase in resolve.

Evidence from the American drone campaign shows similar results. Studying Pakistan from 2007 through 2011, political scientists Patrick Johnston and Anoop Sarbahi find drone strikes correlate with a decrease in both the frequency and lethality of militant attacks.[99] This suggests the drone campaign in Pakistan has reduced militants' capacity, and the negative reaction among Pakistanis is insufficient to replenish it. If the drone strikes increased the resolve of people in the Federally Administered Tribal Areas, if anger over the attacks effectively created more insurgents than the strikes eliminated, or if new insurgents were as capable of executing attacks as the people they replaced, then we would expect to see the opposite result from a study like Johnston and Sarbahi's.

Though limited, the available evidence suggests the drone campaign degraded Al Qaeda's capabilities and contributed to the decline of Al Qaeda Central. Drones have killed approximately 60 senior Al Qaeda figures who provided leadership in operational planning, finances, logistics, communications, and training.[100] Letters discovered in the raid on Osama bin Laden's compound showed Al Qaeda's leader was concerned with the effect of losing these senior members, worrying that when "experienced leadership dies, this would lead to the rise of lower leaders who are not as experienced as the former leaders and this would lead to the repeat of mistakes."[101] Due to the importance of experienced leadership, Al Qaeda is devoting more of its limited time, brainpower, and finances to keeping senior members safe, which means devoting less to developing strategy, crafting propaganda, and planning attacks.

Furthermore, fear of drones has hindered Al Qaeda's ability to meet, communicate, train, and recruit. A document discovered in Timbuktu, Mali, in buildings previously occupied by members of Al Qaeda in the Islamic Maghreb, recommended 22 tactics militants should use to counter the threat of drones.[102] Some are overly ambitious, such as using "a group of skilled snipers to hunt the drone." While they may be able to shoot down smaller, slower UAVs flying closer to the ground, it would be hard to knock down a Reaper with a rifle. Other recommendations underestimate ISR drones' variety of sensors, such as suggestions to spread "reflective pieces of glass on a car or the roof of the building" and form "fake gatherings such as using dolls and statues to be placed outside false ditches to mislead the enemy."

However, some are sensible, such as recommendations to not "use permanent headquarters," avoid "gathering in open areas," hide "from being directly or indirectly spotted, especially at night"—which demonstrates awareness of UAVs' infrared cameras—and take cover under thick trees or in caves. These measures would make it more difficult for drones to spot and attack militants, but also make it harder for the group to meet and move around. Similarly, the document tells fighters to "disembark vehicles and keep away from them especially when being chased or during combat," and "when discovering that a drone is after a car, leave the car immediately and everyone should go in a different direction because the planes are unable to get everyone." This makes sense, since larger objects, such as cars, are easier to track and fire upon than individuals. However, avoiding vehicles hinders Al Qaeda's movement, reducing fighters' ability to switch positions or flee during combat, and decreasing the amount of territory in which a cell can operate.

Additionally, the document instructs operatives to "maintain complete silence of all wireless contacts" and instructs leaders not to "use any communications equipment." These recommendations demonstrate an awareness of intelligence agencies' electronic eavesdropping capabilities. If Al Qaeda avoids anything but in-person communication, it would make it more difficult to discover the group's members and plans. However, it also limits their effectiveness. If members fear using communications equipment and avoid meeting in person except under specific conditions, Al Qaeda's ability to plan and coordinate activity would suffer.

The Al Qaeda document also includes a recommendation to form "anti-spies groups to look for spies and agents" as part of its anti-drone strategy, which indicates how paranoia over drones limits the terrorist network's cohesion and recruitment efforts. Accounts of the central Al Qaeda organization in Afghanistan and Pakistan indicate the group has greeted new recruits from Europe with less trust and given them less training in recent years, at least partially due to fear they might deliberately or inadvertently provide drone operators with targeting information. Successfully vetted recruits receive less training as well, because signature strikes make Al Qaeda hesitant to maintain training camps, leading them to spend much of their time changing location and hiding in caves or small huts.[103] This compounds the problem of leadership losses by reducing the interactions that cultivate interpersonal trust, and hinders remaining leaders' ability to evaluate and promote new talent.

From the perspective of Western governments and security institutions, the most important piece of evidence is Al Qaeda's inability to execute a terrorist attack in the West since 2005. Al Qaeda Central planned and carried out the September 11th attacks in 2001, while Al Qaeda-linked operatives executed the 2004 Madrid train bombings and 2005 London transportation bombings. Since then, the United States has endured relatively small attacks by self-starters, such as the Fort Hood shooting (2009), Boston Marathon bombing (2013), and San Bernardino shooting (2015), but nothing large scale. Europe also faced some attacks, but the only terrorism in Europe since the London bombings that killed 25 or more were committed by a far right nationalist (Anders Breivik in Norway in July 2011: 77 killed) or Al Qaeda's jihadist rival ISIS (operatives in Paris in November 2015: 130 killed; in Brussels in March 2016: 32 killed; and a self-starter driving a truck through a crowd in Nice in July 2016: 86 killed). Based on this track record, it appears the West has largely handled the threat from Al Qaeda; at least for now.

However, it is difficult to determine how much credit drone strikes deserve for this success. Anti-terrorism defenses likely played a role. Geopolitics may have as well, as international jihadists focused more on Afghanistan and Iraq, and later Syria, Yemen, and North Africa.

While decapitation strikes have weakened Al Qaeda, removing leaders and frightening those who remain, some of the most successful attacks came by non-drone means. For example, a joint CIA–ISI operation captured Khaled Sheikh Mohammed in Pakistan in March 2003, and he remains incarcerated at Guantanamo Bay. Most famously, United States Navy SEALs killed Osama bin Laden in a ground raid on his compound in Abbottabad, Pakistan in May 2011.[104] This leaves Al Qaeda Central with few established senior leaders besides Ayman al Zawahiri, a founding member who assumed control after bin Laden's death, but lacks his predecessor's charisma.[105]

This suggests the importance of individual leaders determines the value of particular decapitation strikes. Bin Laden brought together disparate members of Al Qaeda, with many swearing personal allegiance to him, and served as the group's chief spokesman and inspirational figure.[106] Given his centrality to Al Qaeda, he is difficult to replace, and Zawahiri's weaker interpersonal skills and speaking ability may make him incapable of duplicating bin Laden's contributions. By contrast, Hassan Nasrallah proved more than capable of taking over Hezbollah after Abbas al Musawi's death. With this in mind, perhaps the reason decapitation strikes hindered Al Qaeda Central while AQAP remains resilient despite the death of its leader, is Qasim al Raymi is a suitable replacement for Nasir al Wuhayshi. As the longtime head of military operations and a founding member, al Raymi may be able to lead AQAP as well as, or better than al Wuhayshi, especially given the group's focal shift to expanding territorial control in Yemen.

The available evidence, especially the case of AQAP, strongly suggests drone strikes alone are insufficient to defeat a terrorist or insurgent group, though they can play a useful role as part of a larger strategy. Widespread international efforts to track and freeze finances and shut down fundraising fronts have substantially reduced Al Qaeda's operating budget.[107] Heightened and more coordinated intelligence efforts have hindered recruitment, especially in Europe, as has competition from ISIS. Though these countermeasures affect the larger Al Qaeda network, they have focused more on Al Qaeda Central. While AQAP enjoys relative freedom of movement in Yemen, with the most powerful intervening state, Saudi Arabia, focused on their enemies, Al Qaeda Central lost its sanctuary in Afghanistan, and faces continued pressure from Pakistani, Afghan, American, British, and allied forces. Drone strikes and other decapitation efforts have targeted both branches, but have likely done more damage to the leadership of Al Qaeda Central than its Yemen-based affiliate.

The problem is the concerted effort against Al Qaeda requires resources and international cooperation; and nations may be unwilling to devote as much to other terrorist groups. Even at its peak in the late 1990s/early 2000s, Al Qaeda Central was fairly small. Exact figures are difficult to come by, but most put the core group at a few hundred, with some estimates ranging as high as a few thousand.[108] As September 11th demonstrated, a small group can have a large impact. But Al Qaeda Central's size, need for secrecy, and reliance on a few leaders, especially bin Laden, made the group vulnerable to a multifaceted campaign that included drone strikes.

On their own, drone strikes probably cannot defeat a terrorist group, especially a larger one that enjoys some popular support, but they can reduce a network's power and reach, thereby facilitating a relatively inexpensive counterterrorist strategy. Terrorist and insurgent groups that lose leaders and trained operatives have fewer capable members. This can also reduce group cohesion by sowing distrust, and harm operational capacity by forcing the organization to devote more resources to avoiding attacks. As the example of AQAP demonstrates, a group that loses leaders may still make gains, but that does not mean it has an endless stable of talented replacements. Drone strikes almost certainly create some blowback, but the available evidence indicates they also reduce the lethality and frequency of attacks, at least in the short term.

This suggests there is a time lag between popular blowback and an increase in militant groups' capabilities. Anger over drone strikes may help recruitment, but it takes time to train and organize recruits before they can fully replace or exceed the group's losses. Similarly, it takes time for leaders to establish interpersonal trust, and skilled leaders are uncommon. Therefore, a steady campaign of drone strikes may be able to reduce a terrorist group's capabilities and maintain that reduced level, a strategy crudely analogous to "mowing the grass." The grass will grow back, but occasional mowing can prevent it from growing beyond an unacceptable level.

However, one can easily imagine a level of violence that sparks a rapid, powerful blowback, creating a flood of recruits, inspiring a wave of self-starters, and prompting states to support the militant group with diplomatic pressure, material assistance, or even direct intervention. A grass mower can maintain a lawn, but cannot clear brush. Along these lines, a drone campaign may be able to keep the threat from a smaller terrorist group acceptably low, but not stop a larger insurgency. For example, despite facing years of drone strikes, the Taliban, Haqqani network, and other interrelated insurgent groups based along the Afghanistan–Pakistan border continue to operate and execute attacks in both countries. They enjoy some popular legitimacy in the FATA, in part due to Afghan and Pakistani ethnic and tribal politics, which drone strikes cannot counter, even when joined by some counterinsurgency efforts on the ground. A similar problem may apply to Al Qaeda in the Arabian Peninsula. While decapitation strikes may reduce the group's ability to attack distant foreign targets, they probably cannot defeat AQAP as long as it enjoys some popular legitimacy and can take advantage of the Yemeni civil war.

Even when successfully utilized against a smaller organization, "mowing the grass" is unlikely to yield a lasting solution. The hope is enough pressure applied over enough time will make a group collapse, as frustrated members quit and recruits dry up due to a lack of success and fear of attack. However, if blowback continues to boost recruitment and resolve, the strategy has no endpoint—just a perpetual cycle of mowing and re-growing.

A strategy built around drone strikes therefore has limited value, but it may be the best option under resource constraints. The United States and other powerful countries are rarely willing to execute a sustained counterinsurgency campaign. Even if they are willing to at first, as soldiers die and costs accumulate their political will decreases, as per the goals of insurgent strategy. If states perceive a threat from a terrorist group, they could choose not to attack and focus entirely on anti-terrorism protections. However, if they are going to include an element of offense in their strategy, drone strikes put less of their personnel at risk, and may prompt less blowback than alternative options.

To get the most out of a drone strike strategy, counterterrorists need to balance the conflicting goals of reducing terrorists' capabilities and minimizing blowback. This requires avoiding civilian casualties as much as possible, and may benefit from greater transparency. Given the minimal risk to military personnel, the relatively low cost, and the domestic political pressure to employ offensive measures against terrorist groups, drone strikes will remain a factor in asymmetric warfare for the foreseeable future.

CHAPTER 4

Terrorists and Insurgents, Armed with Drones

Around 3 am on January 26, 2015, a small quadcopter crashed on the White House lawn. No one was hurt, and the drone's operator, who worked at the National-Geospatial Intelligence Agency, turned himself in and admitted he was drinking and lost control of the aircraft while flying it for fun.[1] Though this incident caused no damage and appears accidental, it highlights how unprepared the United States is for the security risks associated with drone proliferation.

The Phantom model UAV that crashed near the White House is inexpensive, easy to operate, and difficult to track or prevent from entering restricted airspace. Built by the Chinese company DJI Innovations and marketed globally, the Phantom can carry a small payload and is popular with aerial photographers.[2] The most advanced model costs up to $1,700 but earlier versions can be purchased for less than $500.[3] Though unauthorized aircraft of any kind cannot legally fly above the White House, and operating a drone anywhere in Washington DC is outlawed, the Phantom was too small and flying too low to register on White House radar. An officer on the ground spotted it, but could not prevent it from entering White House airspace or bring it down.[4]

Though this incident was innocuous, a terrorist could easily use the Phantom, or a similar UAV, for something more nefarious. The Phantom can lift payloads weighing up to 600 grams, allowing it to carry a variety of cameras. To put that in perspective, the M67 hand grenade used by the United States military weighs 397 grams and has an effective casualty radius of five meters,[5] which means the Phantom

FIGURE 4.1   DJI Phantom.

could carry a deadly explosive charge. Alternatively, a quadcopter could make for a useful surveillance tool, photographing sensitive locations, such as a bridge, tunnel, or government building. As detailed below, there are multiple known incidents involving organized groups or disgruntled individuals using small drones to monitor or attack various targets in the United States, Israel, and elsewhere.

In the coming years, it will become progressively easier to acquire unmanned weapons and information-gathering systems. Drones appeal to terrorists and insurgents for the same reasons they appeal to states: as a method of acquiring information or striking targets without risking personnel. Given their limited budgets, militant groups probably will not develop their own squadrons of advanced UAVs. Additionally, despite pop culture-driven fears, hacking drones is difficult. Terrorists may be able to tap into UAV video signals or confuse unmanned aircraft, but will probably not take over military drones and order them to fire their weapons. However, with more and more countries producing and selling military robotics technology, it becomes increasingly likely militants could purchase some through the black market or receive some from state sponsors, much as they acquire firearms or explosives.

While many of the most notable developments thus far have been military in nature, the robotics revolution features commercially available automated systems as well. In the early twenty-first century, numerous non-military versions of unmanned aircraft have become available for use by individuals and businesses, and, as with earlier inventions like personal computers or cell phones, this trend will likely accelerate. From monitoring traffic, to shooting movies, photographing real estate, or delivering packages, a large variety of privately controlled drones will become increasingly commonplace in developed countries. Inadvertently, this will provide terrorists and insurgents, from individual self-starters to organized networks, with robots they can adapt to enhance their capabilities in asymmetric warfare against their relatively more powerful state opponents.

As drones become common sights, especially in cities, it will become difficult to identify one that presents a threat; and, as the incident at the White House demonstrates, even if a law enforcement officer does spot a drone observing or attacking a target, it would be difficult to stop. Therefore, states should develop a variety of countermeasures, including electronic defenses—outfitting infrastructure, military bases, government offices, police stations and other potential targets with transmitters that overwhelm any UAV's remote control signal and order it to turn around, land, or return to base.

ACQUIRING MILITARY ROBOTS

Attacks from large unmanned aircraft are unlikely to pose a strategic risk to powerful militaries in asymmetric warfare. Given their resource advantage, states have little difficulty maintaining air superiority in conflicts against insurgencies. With the airspace above any active theater monitored by radar, and the location of friendly aircraft known, any drone large enough to carry missiles is unlikely to escape notice. If states detect an unfriendly or unidentified Predator, or a similarly large UAV, flying in airspace they control, they could target it with ground-based air defense systems.[6] Alternatively, interceptor aircraft could engage and destroy the enemy drone, much as an Iraqi MiG shot down an American-operated Predator on an information-gathering mission over Iraq in 2002.[7] States will likely feel few qualms firing on unidentified or potentially hostile unmanned aircraft, as there is no chance they will kill a pilot, and thus less risk of accidentally harming an innocent or creating an international incident.

However, if insurgents acquire smaller drones that fly at low altitude—which are harder to spot with radar—they could pose a threat to ground forces with kamikaze attacks in a manner similar to the small Switchblade UAV used by the United States military.[8] Such robotic non-line-of-sight weapons are more maneuverable and accurate than alternative measures of striking targets at a distance employed by insurgents, such as mortars or crude rockets. Smaller units operating in urban environments would be especially vulnerable to kamikaze drones, since insurgents could direct them from covered positions, using the UAVs' camera to locate their target. With its flight time of 20–40 minutes and effective range of 10 km, a Switchblade could provide urban guerrillas with an effective method of attacking exposed soldiers or unarmored vehicles without revealing their location.

While fortified military areas may be able to defend against kamikaze UAVs with Counter Rocket, Artillery, and Mortar systems, small drones in the hands of terrorist or insurgent groups could prove especially threatening to civilians, including in developed countries. An individual operative could crash a Switchblade or a makeshift equivalent into a populated area, such as a market, causing damage on the scale of a small bomb planted on the ground. Once an attack is in progress it would be difficult to stop, since air defense systems are designed to monitor and shoot down larger aircraft flying higher above the ground, and C-RAM systems would be impractical in densely populated areas due to the possibility of falling debris and the massive cost of trying to protect everywhere at once.

As with other military technology, governments can restrict UAV sales to friendly states. For example, the United States permits General Atomics, which makes the Predator, and other defense contractors to sell exclusively to allied governments, primarily members of NATO. European companies face similar restrictions, as do drone makers in Israel and other non-European US allies. It is unlikely, however, that other countries with emerging UAV manufacturers, such as China, Russia, and Iran, will allow sales only to governments friendly with the United States, or that any purchasers or manufacturers will refrain from reselling or granting drones to militant groups.

In February 2013, General Atomics reached an agreement to sell an undisclosed number of unarmed Predators to the United Arab

Emirates for $197 million, which is the first time an American company sold large drones to a non-NATO ally.[9] Even though this sale includes only the XP model designed for reconnaissance, rather than an MQ outfitted with weapons, it drew some scrutiny because of the possibility the UAE will use the Predators for domestic spying and repression of political dissidents. Additionally, some critics raised the possibility terrorist groups could steal or purchase a Predator from the UAE, though they have not presented any evidence in support of this speculation. The deal received the required approval from both the State Department and Congress in 2015, and General Atomics delivered the drones in early 2017.[10]

While this sale provides further evidence that the number of countries acquiring UAVs continues to expand, it also demonstrates the political barriers and prohibitive monetary cost that make it unlikely terrorists will acquire Predators or other large UAVs, whether legally or illegally. Governments have strong incentives to prevent theft of any weapons they control, and there is no known instance of a terrorist group stealing manned military aircraft. Sales to foreign entities by UAV-manufacturing defense contractors that do their primary business with the United States and American allies require governmental authorization, and any country suspected of unauthorized transfers would likely forfeit the ability to acquire more drones or the parts necessary for maintenance. While UAV manufacturers based in unallied or adversarial countries may sell to different clients, they are also likely to punish unauthorized transfers by cutting off future sales of aircraft and parts. The incentives for states to control military-grade drone technology reduce the risk large UAVs will be sold illegally or stolen.

Even if a state decides to sell large drones it has manufactured or purchased to a terrorist group, or rogue members of a military try to sell some on the black market, the cost is probably too high to be practical. At its high point in the late 1990s/early 2000s, Al Qaeda's annual operating budget was approximately $30 million.[11] Since September 11th, 2001, efforts to track and freeze the funds of Al Qaeda's financiers and various charitable or business fronts by the United States Treasury Department, other governments including the UK and Saudi Arabia, and international bodies such as the Financial Action Task Force,[12] have reduced it. With each Predator costing approximately $4 million and each Reaper costing more than $16 million when legally purchased in bulk by the United States, large UAVs are too expensive for terrorist and insurgent networks, even before accounting for black market premiums.[13]

Spending that much on a large drone would go against the cost-effectiveness at the core of guerrilla strategy. For comparison, the largest attacks by Al Qaeda or its affiliates cost substantially less than a single unarmed Predator. CIA estimates place the cost of the September 11th operation at approximately $500,000, while the 2004 Madrid train bombings cost $70,000, and the 2005 attacks on London's transit system cost only $10,000. According to Stuart A. Levey, the Under Secretary for Terrorism and Financial Intelligence in the United States Treasury from 2004 to 2011, the majority of Al Qaeda's funds go towards training, operatives' salaries, travel and the purchase of travel documents, payments to families of suicide bombers, and bribes for public officials.[14] Therefore, it would make little strategic sense for Al Qaeda to spend so much money to acquire a larger drone, especially since advanced militaries could spot it and shoot it down.

ISIS' peak budget was much larger, as high as $2 billion in 2015,[15] but buying large UAVs would not have been cost effective for them either. There is no known black market selling Predators or similar UAVs, and even if there were, most of ISIS' funds cover salaries for fighters, as well as police and other domestic administrators. The group levied taxes and extracted oil in areas it controlled but its finances are under pressure. The United States, Iraq, Turkey, and others cracked down on black market oil sales, and bombed ISIS' oil infrastructure and transportation routes.[16] Meanwhile, the group exhausted its one-time cash infusion from capturing banks in the large Iraqi city of Mosul. Losing territory in Iraq and Syria, ISIS halved fighters' salaries starting in January 2016, to $200 per month for locals and $400 per month for foreign recruits.[17] With the United States, Russia, Turkey, and the Iraqi and Syrian governments controlling the skies, it would make little sense for ISIS to devote a chunk of its shrinking budget to large UAVs.

Small drones, however, are much cheaper, and may prove attractive to terrorist or insurgent groups. Under contracts signed in 2011, each Switchblade costs approximately $100,000—which includes training and other services—while each observational Wasp drone costs $50,000.[18] Though still expensive, these or similar drones would not break the bank for a well-funded network. More likely, since small UAVs are easier to produce than the large alternatives, cost less, and can be operated by individuals with little training, states may be willing to give them to militant groups they sponsor.

HEZBOLLAH'S STATE-SPONSORED DRONE FORCE

On October 6, 2012, Israel shot down a mid-sized drone in the Negev desert, near the West Bank's southern border. Hezbollah claimed responsibility. According to the group's leader, Hassan Nasrallah, the drone was manufactured in Iran, assembled in Lebanon, and used for "reconnaissance flights inside occupied Palestine."[19] Hezbollah flew Iran-provided UAVs over Israeli territory before, but never this far from its base in Lebanon.

The first incident occurred in late 2004. The unidentified drone model flew about 1,000 feet above the ground, escaping detection by Israeli radar due to its small size and low altitude. An Israeli officer on the ground spotted it near the Lebanese border. The UAV spent five minutes in Israeli airspace, before turning west towards the Mediterranean Sea, where it crashed. The Israeli military interpreted the incursion as a demonstration of Hezbollah's capabilities, and initiated a review to determine how the flight originally escaped notice.[20]

In April 2005, Hezbollah flew an Iranian-made Mersad UAV over northern Israel. The Mersad, also known as a Mohajer (which means "migrant" in Farsi), was first developed in the 1980s by Ghods Aviation, an Iranian company, for reconnaissance in the Iran–Iraq war. In the years since, Ghods built four versions of the Mohajer, the most recent of which is approximately three and half meters long and capable of flying as fast as 135 miles per hour for a short while, with an operational range of approximately 100 miles. The Mohajer-4 underwent a successful flight test in February 2002, and, unlike the original version, includes autopilot, advanced cameras, and the ability to paint targets with a laser for guided munitions.[21] According to a diplomatic cable from the American embassy in Beirut released by WikiLeaks, Iran provided Hezbollah with three Mohajer-4s in 2004 or 2005, one of which was operational at the time of the flight into Israeli airspace. Sources in Syria and southern Lebanon indicated Syrian intelligence assisted with the UAV's launch, which flew over Israel to gather information and demonstrate Hezbollah's growing unmanned capabilities.[22]

In the 2006 war between Israel and Hezbollah, the Israeli Defense Forces shot down three Hezbollah-controlled Ababil UAVs over

Israel.[23] Also built by Ghods, the Ababil (which means "swallow" in Farsi), is slightly under three meters long, but more aerodynamic than the Mohajer. It looks like a missile with wings, and can reach a top speed of 185 mph, with an operational range of 150 miles. Like other small and mid-sized surveillance drones, such as the Raven, the Ababil transmits images to a ground-based control station. However, unlike the Raven, the Ababil launches using a track mounted on the back of a truck, or via a rocket launch system.[24]

While Ghods designed the Ababil primarily for ISR missions, it can carry a single warhead with up to 50 kg of explosives.[25] Of the three Ababils Israel shot down in the 2006 war, at least one held 30 kg of explosive material.[26] Israeli forces recovered the explosives from an Ababil intercepted by an Israeli F-16, and suspected one additional drone of the three they downed was carrying a similar payload, while the remaining Ababil was outfitted exclusively for surveillance.[27] In the years since the 2006 conflict, Hezbollah has reportedly acquired additional Ababils from Iran, some of which carry 45 kg warheads.

Hamas, the Palestinian group that controls the Gaza Strip and enjoys limited Iranian sponsorship, also controls some Ababils, including an advanced version that may be armed. In July 2014, the Qassam Brigades, Hamas' military wing, released images on Twitter showing an Ababil flying over Gaza carrying what looked like four air-to-surface missiles. However, it is unclear whether the missiles are fake, real but inoperable and included as propaganda, or capable of actually firing.[28]

In September 2014, Hezbollah became the first militant group to successfully attack a target using unmanned aircraft. Intervening in the Syrian civil war in support of the government of Bashar al Assad,

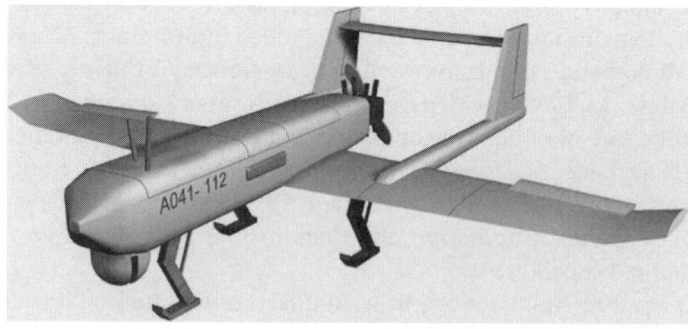

FIGURE 4.2   Mohajer-4, made by Ghods Aviation.

Hezbollah drones bombed a position near the Lebanese border held by Jabhat al Nusra, Al Qaeda's affiliate in Syria.[29] Though reports out of war-wracked Syria are spotty, it appears Hezbollah's drones fired weapons, rather than sacrificing themselves in a kamikaze attack, indicating an increasing sophistication.

Using images from Google Earth, the defense consultancy IHS Jane's spotted a Hezbollah airstrip in Lebanon, about 18 km west of the Syrian border, most likely used for UAV launches. The airstrip, built sometime between February 2013 and June 2014, is too short for most manned aircraft to take off and land, indicating Hezbollah probably does not use it to receive arms shipments from Iran.[30] However, the length could accommodate Ababil reconnaissance drones, as well as the larger, Iranian-made Shahed 129.

The Shahed 129, which Iran revealed in 2012, appears similar to the Predator or the Israeli-made Hermes 450, and is capable of carrying missiles. Larger than the Ababil, the Shahed can remain in the air up to 24 hours and fly up to 2,000 km. Like the UK's similar Watchkeeper UAV, the Shahed offers ISR capabilities, along with the ability to paint targets with a laser to assist missile accuracy.[31] Because Hezbollah's secret airbase could accommodate it, and because it can carry missiles, a Shahed was probably the drone that fired at al Nusra in 2014. Additionally, throughout 2015, Hezbollah used various drones to monitor, designate targets, and bomb al Nusra positions around the contested Syrian town of Qalamoun.[32]

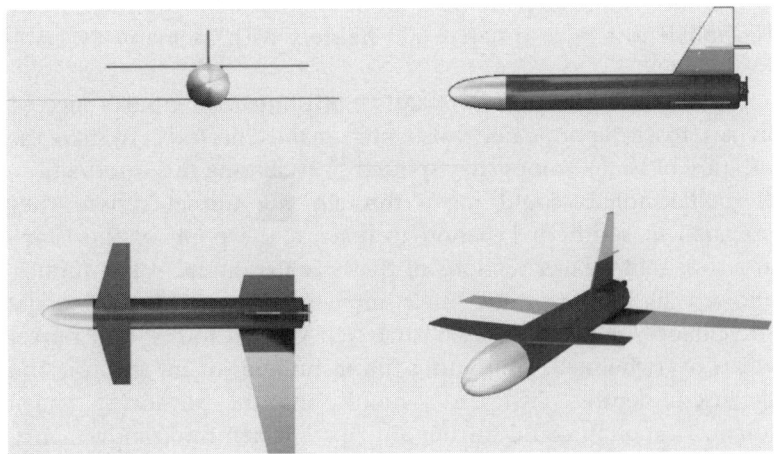

FIGURE 4.3   Drawing of an Abibil UAV, made by Ghods Aviation.

Though Hezbollah's most sophisticated UAV deployments targeted another militant group, these flights demonstrate advancements in robotics can enhance the capabilities of insurgents in asymmetric conflict. With kamikaze or missile-carrying drones providing a more accurate method of attacking state military forces than mortars or rockets, insurgents could create more casualties and deny their enemies an easy victory. In conflicts such as Israel–Hezbollah, where fighting takes place in close proximity to Israel's main territory, UAVs equipped with explosives could threaten nearby civilian populations. While Hezbollah expects to face significant casualties when resisting militarily superior Israel, the same material superiority leads Israelis to expect decisive victory and minimal casualties, especially among civilians inside Israel proper. By improving militants' ability to kill soldiers and civilians, armed UAVs can help them impose additional costs, which may convince their stronger opponents to halt offenses, negotiate ceasefires, or withdraw forces.

Small UAVs' information-gathering capabilities can enhance militant groups' strategies as well. In 2006, Israeli ground forces advancing on Hezbollah-controlled positions in the mountains of southern Lebanon faced fierce resistance. Despite a significant resource advantage, with 30,000 soldiers backed by armored vehicles and aircraft against an estimated 10,000 fighters, Israel was unable to secure southern Lebanon or advance more than a few miles beyond its border. Before withdrawing, the Israeli Defense Forces lost 116 soldiers,[33] with an additional 628 wounded, while Hezbollah lost an estimated 600 fighters with as many as 1,500 wounded.[34]

Hezbollah leader Hassan Nasrallah attributed his group's success, in part, to a cell phone network, which enabled his forces to share the location of Israeli troops they spotted.[35] By sharing this information, Hezbollah forces could move through the tunnel system they prepared in southern Lebanon to mass at the point of the Israeli attack or raid weaker sections of the Israeli columns.[36] The tunnels allowed fighters to hide their movements from Israel's aerial surveillance, pop out to ambush Israeli ground forces, and retreat when overwhelmed, combining the techniques of hit-and-run and defense-in-depth. With UAVs monitoring the advancing Israeli forces, Hezbollah could further anticipate Israeli troop movements, and prepare their defenses accordingly.

As an invading force entering mountainous territory familiar to their opponent, Israeli troops were at an informational disadvantage. Hezbollah scouts could observe the movements of Israeli columns climbing the foothills or moving through passes, while most of their forces remained hidden in tunnels. When Hezbollah fighters launched rockets into Israel from fixed batteries or the backs of trucks, they would reveal their location to Israeli radar and aerial surveillance, after which Israel would attempt to destroy the rocket batteries with targeted missiles. However, Israel could not see the underground movement of Hezbollah guerrillas or differentiate between civilian vehicles and trucks carrying concealed rockets. Hezbollah utilized this informational advantage to shoot almost 4,000 rockets into Israel over the five weeks of the conflict,[37] killing 43 civilians and causing "serious" or "moderate" wounds to an additional 76.[38]

If Israeli forces had been able to secure southern Lebanon, they would have greatly reduced Hezbollah's ability to fire rockets into populated areas of Israel. The maximum range of a Katyusha rocket, which made up the majority of Hezbollah's arsenal in 2006, is approximately 25 km. When fired across the Lebanese border, this limited range makes Katyushas capable of threatening only northern Israel.[39] Major population centers are farther from the Israel–Lebanon border, with Tel Aviv about 100 km away. Israeli airstrikes and ground incursions were able to destroy numerous rocket launchers, but unable to prevent daily rocket fire throughout the conflict. This inability to prevent an ongoing threat to its civilians likely contributed to Israel's willingness to accept a United Nations-brokered ceasefire.

In the years since the 2006 conflict, Hezbollah reportedly acquired missiles with greater range, allowing them to set up batteries further from the Israel–Lebanon border in anticipation of another Israeli ground invasion.[40] In November 2012, at an event marking the Day of Ashura, Hezbollah displayed a Fajr-5 missile it acquired from Iran.[41] The Fajr-5 has a maximum range of 75 km, which, if fired from the Lebanese border, could easily hit Haifa and potentially reach the suburbs of Tel Aviv. It is larger and flies faster than Katyusha rockets, making it more difficult for C-RAM systems, such as Israel's Iron Dome, to shoot down. Even with a robust Iron Dome presence on the Lebanese border, some missiles would likely get through. This increases the importance for Israel of securing territory further into Lebanon to push Hezbollah out of range of its largest population centers and disable the batteries stationed farther from the border.

However, with unmanned aircraft observing Israeli movements while Hezbollah's forces remain hidden until firing, Hezbollah would enjoy a larger informational advantage if Israel attempted to invade southern Lebanon again in the future. Mersad, Ababil or Shahed drones could spot Israeli ground forces at a distance, granting Hezbollah greater lead time to mass forces or move rocket launchers. With Hezbollah forces moving through tunnels or other prepared cover, they would be difficult for Israeli UAVs to spot from the air. Therefore, in the event of another Israel–Hezbollah war, drones have the potential to increase Hezbollah's informational advantage, improving their ability to deny Israel victory.

Similarly, urban insurgents could augment their informational capabilities utilizing small UAVs. Observational drones would allow insurgents to monitor the movements of enemy soldiers, helping them determine the patterns of patrols, improving their ability to set up roadside bombs or plan ambushes. Quadcopters watching the streets near meeting locations, safe houses, bomb-making factories, or weapons caches would provide insurgents with advance notice of raids, granting them a window of opportunity to disperse, or to hide or destroy incriminating material. Small ISR drones could therefore enhance the informational capabilities of both urban and rural insurgencies, such as those fought in Iraq or Afghanistan against American and allied forces.

HACKING UNMANNED SYSTEMS

Along with controlling their own UAVs, terrorist and insurgent networks could hack into opponents' drones. This could involve tapping into video feeds, or confusing drones' navigation systems and causing them to fly off course. However, no one has demonstrated the ability to take over a military drone and launch its weapons, and it is unlikely a terrorist will be able to do so.

In December 2009, the United States admitted insurgents had intercepted the video feeds of Predator drones operating over Iraq. They used software called SkyGrabber—an "offline satellite internet down-loader" designed to gather free-to-air movies, music, and pictures from satellite internet providers[42]—to view the footage as it transmitted from the drone back to base via satellite. The software can be purchased legally for as little as $26, or less than $100 when accompanied by a tuner card that receives satellite transmissions. American forces confirmed they found "days and days and hours and hours" of video taken by Predators on captured insurgent laptops, and that the insurgents had distributed the footage to multiple organizations.[43]

Intercepting a Predator's video feed falls far short of electronically taking control of an MQ drone and firing Hellfire missiles. The insurgent hackers were not able to direct the UAV's flight path or the position of its cameras. However, accessing the video transmission provided the insurgents with valuable information. Not only were they able to see everything captured by the cameras, they also learned which targets the United States was keeping under surveillance. With this knowledge, they could move activity they wanted to keep secret to a different location, act more carefully in places they knew were on camera, or deliberately feed false information to US intelligence analysts.

The Predator video feed was unencrypted, which left it vulnerable to simple software like SkyGrabber. This particular problem, therefore, has a simple fix. However, it reveals a vulnerability in remotely operated UAVs. To navigate and communicate with operators, they rely on signals that travel thousands of miles and bounce off satellites. Encryption can protect against most efforts at corruption or interception, but rendering these signals completely secure 100% of the time is difficult, if not impossible.

In 2012, a research team from the Radionavigation Lab at the University of Texas used a technique called "spoofing" to misdirect an unmanned aircraft. Demonstrating the technique for the Department of Homeland Security on a university-owned drone, the researchers sent a false GPS signal, which caused the drone to fly to a different location than its operators ordered.[44] UAVs, both commercial and military, rely on the Global Positioning System network of satellites to determine their current location and the location of their targets. By sending a "spoofed" GPS signal, the University of Texas team convinced the drone it was in a different place than it actually was. As a result, it altered its flight path to travel from the fake location to the originally programmed destination. This caused the drone to veer off course and fly to a different destination, even while its navigating software believed it was arriving at the pre-programmed target.

Iran may have used a similar spoofing technique to bring down an American RQ-170 Sentinel flying a covert reconnaissance mission over its territory in late 2011. The Iranians claimed they jammed the signal between the drone and its operators, which caused the plane to switch to autopilot. Then, using fake GPS signals, they tricked the Sentinel into landing by making it think it was back at base.[45] The United States has not confirmed this account, but did admit American operators lost control of the drone while it was flying a mission over western Afghanistan near the Iranian border.[46] Shortly thereafter, the Iranians proudly displayed what appeared to be an undamaged Sentinel, which is more consistent with a controlled landing than a crash.

Considering the Sentinel's likely height at the time, and the strength of the American signal, Iran needed a powerful signal to jam the remote control connection and send spoofed GPS coordinates, but the University of Texas researchers redirected their drone using signals produced by commercially available equipment that cost approximately $1,000.[47] Though they overwhelmed a weaker signal, the experiment still demonstrates how insurgents could acquire the means to launch electronic attacks on UAVs. Spoofing circumvents the more difficult technique of hacking into the signal issuing directions to the drone, which would enable the hackers to issue new orders, from new destinations to missile attacks.

Though no one has demonstrated the ability to take over a military UAV, hackers have hijacked commercially available models. For example, two cheaper quadcopters made by Parrot—the Bebop, which sells for $400–$500, and the AR, which costs only $250—are vulnerable to attacks using WiFi signals. A hacker named Samy Kamkar created software called SkyJack, which he programmed into a drone that, as demonstrated in online videos, can "autonomously seek out, hack, and wirelessly take over [Parrot] drones within wifi distance, creating an army of zombie drones under your control."[48]

At the Ethical Hacking Conference in May 2016, computer programmer Mark Szabo took complete control of a Parrot AR using a laptop with normal wireless capabilities.[49] To receive instructions, the AR creates its own wireless network, much like a router, but the network is open, requiring no password. This allows an operator to control it with a smartphone or tablet, but unauthorized devices can connect as well. Szabo connected to the drone, sent it code to disconnect from other devices, and then issued orders while blocking the original operator's smartphone from reconnecting.

The AR is basically a toy, and pricier quadcopters, such as the Phantom, do not appear vulnerable to SkyJack-type WiFi attacks, though they are vulnerable to GPS spoofing. However, in a more concerning example, security researcher Nils Rodday claimed at the 2016 RSA security conference he hijacked a larger, more advanced quadcopter used by police and fire departments.[50] Rodday did not reveal the model and manufacturer, in accordance with a non-disclosure agreement he signed in exchange for a drone to test his theory, but he said it is about three times the size of a Phantom, costs $30,000–$35,000, and can remain in the air for 40 minutes. Unlike the Parrot drones, this one does not use an open wireless network; but it relies on an easily cracked security algorithm called "wired-equivalent privacy" (WEP), and avoids using an encrypted radio protocol to communicate with its operator because the encrypted signal creates lag time. According to Rodday, by exploiting these two vulnerabilities, "you can inject packets and alter waypoints, change data on the flight computer, set a different coming home position. Everything the original operator can do, you can do as well."[51]

These examples highlight two potential risks, as military and commercial drone use expands:

(1) Though Rodday and other security researchers will help manufacturers identify vulnerabilities, resourceful hackers

FIGURE 4.4   A Parrot AR Drone 2.0 flying in Nevada.

could still exploit out-of-date security on older equipment, or create new techniques that hijack a more advanced drone.

(2) Even if signal encryption continues to protect against adversaries gaining control of military drones, there is no alternative to the GPS system, and creating one would be very expensive. Therefore, computer-savvy individuals could use spoofing or a similar technique to trick a military or commercial UAV and make it veer off course, disrupting drone missions, capturing UAVs for sale or study, or, potentially, crashing one into a target.

COMMERCIAL UAVS

Terrorists or insurgent groups probably will not be able to take over military drones electronically and, unlike Hezbollah, most do not enjoy significant state sponsorship, and therefore do not have patrons giving them military UAVs like the Ababil or Shahed. However, even if they are unable to steal military UAVs, buy them on the black market, or receive them from state sponsors, they could adapt non-military drones into weapons or information-gathering systems. Any remotely controlled vehicle with a camera that can transmit real-time video could function as a basic ISR platform. Add some explosive material, and any small robot could act as the crude equivalent of a Switchblade and perform a kamikaze attack.

The commercial drone industry in the United States is in its infancy, and expected to grow dramatically in coming years. Hobbyists could always fly small "recreational" UAVs short distances at low heights under the rules pertaining to model airplanes. However, commercial flights became legal in August 2016, when the Federal Aviation Administration (FAA) introduced a regulatory framework. Before then, businesses required a special license.

Small commercial drones, which the FAA defines as any under 55 pounds "conducting non-recreational operations," are limited to daylight flights within the line of sight of the operator. The drone has to remain below 400 feet, though it can go higher if it is within 400 feet of a structure, allowing pilots to fly UAVs above a bridge or building. They cannot fly faster than 100 mph, and have to stay out of airport flight paths and restricted airspace. Operators require an FAA-issued remote pilot certificate.[52]

Before the United States opened the skies to business drones in August 2016, commercial UAV flight was legal in many economically developed and developing countries, including the UK, France, Germany, Japan, China, and Australia. Regulations vary by country, but all adhere to similar principles: looser restrictions regarding smaller UAVs, with many governments requiring licenses to operate larger drones; height restrictions, typically 122–150 meters (about 400–500 feet); requirements operators keep drones within their line of sight, and/or within a specified horizontal distance; bans on flying near airports; and limits on flying in populated areas.[53] Brazil's laws are more restrictive, requiring drone operators to get authorization at least 15 days before intended use, but the government has found this difficult to enforce, especially during the 2016 Summer Olympics in Rio.[54] Russia's are relatively restrictive as well, treating UAVs as regular aircraft in most areas, and banning them outright in Moscow.[55] India is the largest economy to ban drone flight, regardless of intended use. Legally flying a UAV in India requires special permission.[56]

Despite regulatory delays, the UAV market has grown rapidly. In 2014, worldwide drone spending totaled $6.4 billion, and most analysts expect dramatic increases in the near future.[57] Nearly 90% of worldwide sales were military, but the civil side is expected to grow faster, perhaps 20% annual growth over the next decade, compared to 5% for military drones.[58] Investment bank Goldman Sachs forecasts $100 billion in sales between 2016 and 2020: $70 bn military, $17 bn consumer, and $13 bn commercial.[59] According to FAA estimates, as many as 30,000 private and government drones could be legally flying over the United States by 2020.[60] Though predictions vary, all expect the UAV market to grow, which also means it will be easier for terrorists and insurgents to acquire drones and incorporate them into their strategies.

Small UAVs, ranging in cost from less than $100 to almost $1 million, have proliferated in the early twenty-first century. Like the comparatively complex and expensive military UAVs, many of these commercially available drones gather information. The most common is a small body held aloft by multiple rotaries known as a "multicopter," most of which include, or can carry, a small camera. These drones photograph properties for real estate agents, take pictures or videos of celebrities for paparazzi, monitor livestock while feeding live images to farmers through a wireless connection, and much more. Larger and more expensive commercial drones combine video cameras with infrared sensors, and are used by a variety of organizations, from law enforcement agencies gathering surveillance or searching for missing persons, to oil companies tracking spills. Non-informational uses include crop dusting, managing road traffic after accidents, and dropping water on wildfires.[61]

One of the biggest proponents of fully legalizing commercial drone flights is the film industry. Hollywood's primary lobbying group, the Motion Picture Association of America (MPAA), first disclosed in October 2012 that it had been pushing the FAA to allow filmmakers to use unmanned aircraft in United States airspace.[62] The industry plans to use small UAVs to shoot TV and movies from the air, replacing current methods that are more expensive, more restrictive, and more dangerous. According to MPAA spokesman Howard Gantman, cameras on small unmanned aircraft would enable innovative camera angles, and "could be used much more safely than going up a tree and much more cheaply than renting a helicopter."[63] For comparison, studios capturing footage from the air typically rent helicopters for $1,700 per hour, plus an additional $1,900 per day for a pilot, while getting the same shots from a drone could cost less than $1,000.[64]

One option comes from the Belgian company Flying-Cam, which leases an unmanned aerial vehicle called the Special Aerial Response Automatic Helicopter, or SARAH. The SARAH weighs 55lbs, can take off and land vertically, and remain in the air for an hour. Designed for commercial filming, it includes a stabilized "Gyro Head" that carries a high-resolution digital camera, and can record video or broadcast live.[65] In 2014, the SARAH won Flying-Cam an Oscar, the "Science and Engineering Award."[66] Eon Productions, a UK-based film company, utilized a SARAH to shoot some footage in Istanbul in 2012 for the James Bond film *Skyfall*. The drone followed alongside 007 as he chased after a train on a motorcycle, all while maintaining a steady horizon.[67] In October 2014, the FAA granted Flying-Cam one of its first Section 333 exemptions for commercial flight in the United States.[68]

While these innovations help businesses, they also create security risks. There is little effective difference between drones designed to help movie directors capture aerial shots and information-gathering UAVs built for military purposes. The SARAH is bigger than the Black Hornet miniature helicopters used by British soldiers in Afghanistan, and the two drones use different cameras, but both are ISR platforms. An insurgent could acquire a UAV designed for commercial filming and use it to observe soldiers' movements or scout locations for attacks. The broadcast feature would enable real-time monitoring of a given area, while the recording feature would allow insurgents to film potential targets and study their security to design attacks with greater chances of success.

While aerial filming may be an obvious use for unmanned aircraft, there are other, less apparent commercial uses, such as food delivery. In late 2012, researchers at Darwin Aerospace in San Francisco designed the Burrito Bomber, a small unmanned plane capable of dropping an item—in this case Mexican food—via parachute to a pre-programmed target.[69] The engineers at Darwin Aerospace got the idea from the conceptual Taco Copter, a Mexican-food delivery multicopter that attracted attention on the internet in early 2012, but was never actually built. John Boiles, one of the designers, explained they focused on burrito delivery because "Mexican food is really popular" and "burritos are kind of bomb-shaped."[70] Darwin Aerospace has produced demonstration videos and designed a smartphone app for customers to order burritos, but is still awaiting FAA authorization.[71]

The Burrito Bomber may sound ridiculous, and its parachute idea may not be feasible—how would it avoid snagging on tree limbs or power lines?—but delivery-by-drone is likely in the future. The economic incentives are huge, especially for a large company. Much as shooting movies with UAVs is a lot cheaper than hiring a helicopter pilot to carry around camera operators, delivering packages via automated drone would be considerably cheaper than deploying delivery drivers. Automated package delivery would violate most countries' requirement that drones remain in their operators' line of sight, but companies will keep lobbying and, once they have satisfactorily addressed safety concerns, legalization will follow.

At the forefront of this effort is Amazon, the online retail giant. In 2013, Amazon CEO Jeff Bezos announced the company aimed to deliver packages weighing five pounds or less by "octocopter" drones in as soon as five years.[72] Considering the legal restrictions in the United States, Bezos' announcement was at least partially a publicity stunt, but the company continued pushing forward with what it calls "Amazon Prime Air." In March 2015, Amazon tested octocopter delivery systems in rural sites in Canada, with permission from the country's airspace regulator Transport Canada.[73] The octocopters can place a package directly on the ground, avoiding the problems of a parachute delivery system. In December 2016, the company completed its first successful drove delivery in a Prime Air trial program consisting of two customers in Cambridge, England.[74]

In November 2015, Amazon revealed a new helicopter–plane hybrid UAV designed for package delivery. Demonstration videos show the drone, which weighs slightly under 55 pounds, using rotaries to take off vertically, like a multicopter, and then a propeller to fly forward, like a fixed-wing airplane.[75] As a result, it can fly up to 400 feet vertically and 15 miles horizontally, and land smoothly to place packages, carried in an underside compartment, on the ground.[76]

Amazon also filed for a patent on a proprietary "sense-and-avoid" system that uses laser, sonar, and other sensors to perform the tasks its drones would need to deliver packages, such as negotiating obstacles and landing. Gur Kimchi, the head of Amazon Prime Air and one of the names listed on the sense-and-avoid patent application, advocates for a "drone highway" between altitudes of 200 and 400 feet to facilitate delivery and other commercial activity.[77] This would reduce the risk of crashes with manned aircraft in urban airspace, such as police and media helicopters, which typically fly above 500 feet.

Given the potential economic benefits, national regulators will almost certainly expand authorizations for commercial drone flight, perhaps within a designated segment of airspace, which will create large opportunities for Amazon and other companies, and new difficulties for security services. Many anti-terrorism defenses depend on spotting something out of place. Police, security guards, customs officials and other security personnel are trained to identify odd and potentially dangerous behavior, such as someone abandoning a backpack. To help law enforcement officers, New York's Metropolitan Transit Authority started telling passengers in 2002 "if you see something, say something," and dozens of transit and governmental organizations in multiple countries adapted variations of that slogan.[78] This type of vigilance can spot an out-of-place drone now, when flights are uncommon, but will become increasingly difficult as commercial flights proliferate.

SMALL, COMMERCIALLY AVAILABLE DRONES AND TERRORIST STRATEGY

Like many other technologies, as UAVs become more ubiquitous, they will become cheaper and easier to acquire. As with computers, cell phones, and other information technology, commercially available drones will end up enhancing the capabilities of terrorist and insurgent networks. Insurgent organizations could use information-gathering drones to monitor counterinsurgent troop movements, helping them set up ambushes or avoid raids. Terrorists and saboteurs could scout the security of potential targets to determine the ideal time and location to strike. Or they could simply load a drone with explosives and fly it into something. This would fulfill a similar function as a car bomb, but it could fly over barriers, and would not require sacrificing a driver. Unfortunately, it would not be surprising if, in the next decade or two, a terrorist loaded a commercial drone with explosive material and tried to crash it into a building, bridge, or crowded area in the United States or another economically developed country.

The FBI has already thwarted one such attack in the planning stage. In July 2012, Rezwan Ferdaus, an American citizen who was born in Massachusetts and received a degree in physics from Northeastern University, pleaded guilty to charges of attempting to destroy or damage a federal building and providing material support to terrorists.[79] Ferdaus was arrested in September 2011 after outlining his plan—in which he intended to crash drones loaded with explosives into United States landmarks including the Pentagon and the Capitol building—to FBI agents posing as Al Qaeda operatives, and accepting delivery of hand grenades, AK-47s, and C-4 plastic explosives. Ferdaus had designed and built cell phone-triggered detonators, obtained a remote-controlled replica of an F-86

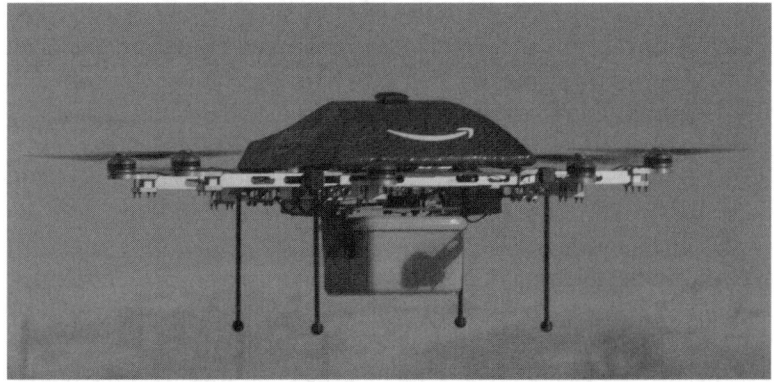

FIGURE 4.5   An Amazon octocopter delivery drone.

FIGURE 4.6   Amazon's next generation delivery drone.

Sabre using a false name, and scouted locations in Washington DC from which to launch the planes.[80] Modeled after the 1950s-era fighter jet, the F-86 replica is almost three feet long, requires extensive assembly, and retails for under $200.[81] Though authorities stopped this planned attack before it advanced to the execution stage, it presages the possibility of similar attempts in the future.

The type of remote-control model airplane Ferdaus planned to use has been available for decades, though there is no publicly known case of someone previously attempting to employ one in a terrorist attack. This suggests he got the idea from news reports of American drone strikes in Afghanistan, Iraq, Pakistan, and Yemen. According to the federal affidavit, Ferdaus was obsessed with using unmanned planes for an attack inside the United States, and saw himself as a devoted member of the global jihadist movement.[82] He had been under FBI surveillance since 2010, when he attempted to supply Iraqi insurgents with homemade cell phone detonators for IED attacks against American soldiers.[83] Like many other self-starters, Ferdaus was not a member of a terrorist group, but frequented jihadist websites, and claimed discussions on those forums helped him realize America is "evil" and that violent attacks against Federal targets in Washington DC could be his contribution to the "solution."[84]

He may have seen a kamikaze UAV attack as quid pro quo for the American drone campaign, or perhaps just thought it would be the most effective method to deliver explosives.

After it became public, Ferdaus' plan was mocked on the DIY Drones internet forum, which calls itself "the leading community for personal UAVs." One member pointed out the model F-86 Ferdaus planned to use requires "a substantial dedicated runway, and plenty of flying practice," which means there was a decent chance he would have crashed while trying to take off.[85] Others expressed relief Ferdaus selected older model airplanes—rather than more modern personal-use drones that can carry larger payloads and be easily adapted to autonomous flight—fearing the government would crack down on their hobby.[86]

The existence of this do-it-yourself drone community indicates the extensive information on UAV construction and modification available online. For example, the DIY Drones website offers instructions on how to build an "amateur UAV" from parts that retail for a few hundred dollars. Whether plane, helicopter, or multicopter, DIY Drones defines a UAV as "an aircraft capable of autonomous flight, without a pilot in control."[87]

To expand access to this capability, the DIY Drones community created ArduPilot, a small, dedicated computer chip that enables autonomous flight for UAVs. Billed as "the world's first universal autopilot,"[88] ArduPilot is based on the Arduino open source electronics platform, a single-board microcontroller that was first released in 2005 and retails for under $25.[89] As an open source platform, the Arduino software can be downloaded for free, and runs on Windows, Mac OS X, and Linux.[90] The latest version of the Arduino chip pre-programmed with the autopilot software, known as ArduPilot Mega, retails for about $185. It includes gyros for controlling balance, pressure sensors, and a GPS system to assist with navigation. The mission planner software is free, and allows users to program predetermined flight paths and analyze mission logs afterward.[91]

This means anyone with a modicum of technical savvy and the ability to perform a simple internet search can find their way to the DIY Drones website and learn how to construct a small UAV capable of autonomous flight. The parts and software are inexpensive and available for purchase on a variety of websites, recreational flights by individuals are legal, and DIY Drones forums are open to all. While the website operators and active participants all appear to be well-intentioned techno-hobbyists and students, there is nothing preventing terrorists from utilizing the information.

Official DIY Drones policy bans any discussion of "military or weaponized applications" or "illegal or harmful use of UAVs;" and the community has "encouraged all relevant regulators, defense agencies and law enforcement agencies to become members" to help them "understand what's possible with amateur UAVs, so they can make better-informed policies and laws."[92] However, it would be easy for someone to use DIY Drones to assist with UAV construction and operation without informing the community of illegal or harmful intentions. Acknowledging the possibility some participants may fail to follow the community's policies, the DIY Drones mission statement declares "we follow the current interpretation of the FAA guidelines" on recreational UAV use, but if anyone does not, "we're going to assume you've got the proper FAA clearance or we don't want to know about it."[93]

DIY Drones' most popular section focuses on multicopters, which users can adapt for autonomous flight using a variation of ArduPilot called ArduCopter.[94] Given their size and payload capacity, the Phantom, Amazon's octocopter, and other small multicopters cannot transport enough explosive material to destroy a building or knock down a bridge, though they could carry the equivalent of a grenade. However, multicopters might be most useful to terrorists as information-gathering platforms.

A quadcopter with a camera could help a terrorist case a target, recording information about the structure, or the presence of security. For example, New York City bans photography near the entrances of tunnels, and on or close to bridges, to prevent anyone from taking pictures they could use to search for structural flaws or any other information they could exploit in an attack.[95] Many cities in the United States, UK, and other developed countries have similar restrictions for infrastructure and government buildings. If someone takes a photo or records video in these restricted areas, the police may confiscate their equipment.[96] However, a multicopter filming potential targets would be more difficult to notice or confiscate. Even if a police officer observes one in a restricted area and manages to bring it down, the operator could remain unknown, because the officer would not have a face-to-face encounter with the photographer. This would be even more likely in the case of multicopters adapted for autonomous flight.

In addition to photographing potential terrorist targets, multi-copters equipped with cameras could help insurgents in active conflicts gather information on locations they plan to attack. By photographing or recording video of police stations, government offices, and military installations, insurgents could determine the ideal time to strike with a raid or a car bomb, using the information they acquire to discover when security patterns offer windows of opportunity. This would provide them with an alternative to scouting targets in person, reducing the risk of getting noticed by police, guards, or security cameras.

As commercially available drone technology develops, more companies are producing systems that could double as ISR platforms. The Solo, a quadcopter from 3D Robotics around the same size as a Phantom, advertises as "the world's first smart drone" and comes equipped with automated flight features. Designed especially for aerial filming to help amateurs acquire professional-seeming footage, the Solo can automatically circle a subject to capture a 360-degree wrap-around shot, follow and film a designated person or object, and move back and forth along a preprogrammed fight path. It takes off on its own with the push of a button, uses a video game-style controller, and can stream live high definition video to any smart phone or tablet from up to half a mile away. The full system with the most advanced camera equipment retails for less than $2,000.[97]

These sort of small UAVs have limited battery life—for example, the Solo can fly and film for about 20 minutes per charge—and therefore could not provide sustained surveillance, unlike more sophisticated military drones. However, they could still give terrorists and insurgents the ability to observe potential targets at little risk. Government security services could shoot down small drones, but with the ability to stream live video, insurgents could still gain valuable information before losing their UAVs. Additionally, because of the low cost, militant groups could afford a fleet of quadcopters, cycling through multiple drones to maintain longer coverage.

ISIS' DRONE BRIGADES

No terrorist group has done more to adapt inexpensive UAVs for military purposes, especially surveillance and propaganda, than the Islamic State of Iraq and Syria. In August 2014, ISIS posted a video

FIGURE 4.7   3D Robotics Solo.

online that included some aerial footage of the Taqba airbase base in Syria, which the group had recently captured from the Syrian government. During the video, viewers can hear militants planning their assault on the base, with one suggesting a target for a truck bomb that would open "the way so that a second suicide bomber can hit the headquarters."[98] Later analysis indicated the footage came from a camera mounted on a Phantom quadcopter.[99]

ISIS also used drone surveillance to coordinate its fighters' activities and spot enemy positions in an assault on the Baiji oil refinery complex, Iraq's largest, as well as in battles near Fallujah, Zawbaa and other Iraqi cities.[100] In 2015, ISIS released a series of videos featuring drone footage, along with shots of commanders monitoring the UAVs' video feeds and directing fighters to the locations of Iraqi artillery pieces. Given the Phantom and similar models' short battery life, ISIS likely employed multiple drones, rotating new ones to the battlefield to maintain coverage or spot new activity.

By showcasing the new technology and ISIS' ability to coordinate sophisticated operations, these videos encourage ISIS fighters and potentially frighten its enemies, while also serving the group's recruitment efforts. In addition to aerial footage matched with battle planning and tactical coordination, ISIS released a slick propaganda video that looks similar to military video games like Call of Duty, featuring drone-shot footage of fighting in the Syrian–Kurdish town of Kobane near the Turkish border.[101] By glorifying combat, and drawing parallels to video games, ISIS aims to recruit young men from around the world.

More recently, ISIS began outfitting drones for attack. In October 2016, Kurdish fighters shot down a small fixed-wing UAV and transported it back to base. When they tried to take it apart, it exploded, killing two. The bomb looked like a battery, suggesting ISIS may have designed the drone to kill in this manner—not in the midst of battle, but after someone shot it down and investigated. This was the first time an ISIS drone killed anyone, but not the first attempt. A few days before the exploding battery killed two Iraqis Kurds, an ISIS-controlled UAV attempted a kamikaze attack, crashing into an Iraqi checkpoint. It exploded, but did not hurt anyone.[102]

In the following months, ISIS expanded its drone attack capabilities, utilizing both fixed-wing and multicopter UAVs. In January 2017, United States Special Operations Forces assisting Iraqis trying to retake Mosul reported attacks from small unmanned planes, about six-feet wide. Instead of kamikaze attacks, these drones carried a small explosive, which they released over a target. Though too small to transport a larger bomb or rocket, they can carry mortar shells with a blast radius of 10 meters, enabling considerable damage. ISIS claimed its newly established "unmanned aircraft of the mujahideen" unit killed or wounded over 30 Iraqi soldiers in its first week.[103]

That figure is probably exaggerated, but the rapid increase in ISIS' drone capabilities concerned its opponents. American aircraft,[104] Kurdish Peshmerga ground forces,[105] and the Syrian army's air defenses[106] have all shot down small ISIS-operated UAVs. However, quadcopters and small unmanned airplanes outfitted for attack may escape notice until it is too late, and observational drones often remain far enough away that it would be difficult to down them with small arms fire from the ground. Counter Rocket, Artillery, Mortar systems could shoot them down, but most C-RAMs are not portable, and mobile variants are expensive.

As ISIS demonstrates, militant groups can amass many small, cheap drones, but their ability to acquire sophisticated UAVs to strike targets or provide sustained reconnaissance is limited without the sort of state sponsorship Hezbollah enjoys. However, ISIS' robotic capabilities indicate a new problem: a fleet of attack drones that overwhelm countermeasures. For just $50,000, insurgents could acquire dozens of UAVs and outfit them with explosives. Based on the American-backed coalition's experience in Iraq and Syria in early 2017, state militaries do not yet have the means to counter it.

THE THREAT TO CIVILIAN AIRCRAFT

Privately operated UAVs also pose a danger to civilian aircraft, presenting a challenge for national airspace regulators. As more commercial and personal drones take to the sky, the chances increase that one will accidentally or deliberately collide with a commercial airliner, private plane, or helicopter. In 2015, there were many more incidents than in previous years—including midair near misses and small drones entering the restricted airspace above airports—and regulators are struggling to keep up.

As the following graph shows, America's Federal Aviation Administration has recorded progressively more incidents involving unmanned aircraft systems. This data covers February 1, 2014 through January 31, 2016, and includes reports from pilots, air traffic control towers, and a few from spotters on the ground.[107] Almost all took place where UAV flight is illegal, including over airports and populated areas. Among the most concerning were a few cases in which a medevac helicopter had to delay takeoff from a hospital helipad to avoid the risk of collision with a nearby drone. This list only includes incidents reported to the FAA, and is therefore surely undercounting the total.

The monthly average in 2014 was 18.58, compared to 92.75 incidents per month in 2015. In both 2014 and 2015, there are more incidents in warmer months, which suggests people are more likely to fly drones when the weather is nice. Nevertheless, every month in 2015 featured more incidents than the same month the previous year, showing a clear upward trend.

Airport incidents followed a similar pattern. This includes every report in which a UAV was spotted within three miles of an airport or airfield, or within five miles of a major airport—such as New York's LaGuardia, Chicago's O'Hare, or Los Angeles International—and in an established flight path. In some cases, control towers reported small drones flying over, or just off the end of, runways. For example,

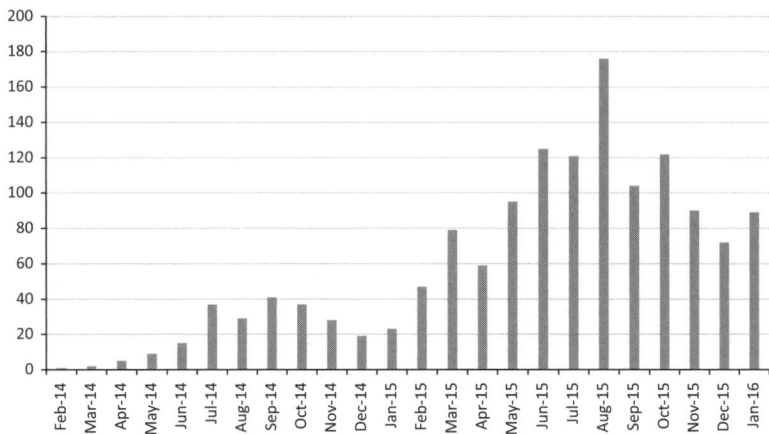

GRAPH 4.1   UAS Incidents Reported to FAA.

in September 2015, Dallas' Love Field airport observed a UAV flying 300 feet above a runway and had to divert all air traffic. These flights took place in restricted airspace, demonstrating regulators' inability to prevent drones from flying where they are not allowed.

There were also dozens of close calls with aircraft, in which a UAV came within 50 feet of a plane or helicopter, or pilots reported taking evasive action to avoid colliding with a drone. Perhaps the most concerning were near misses in which the pilots did not take evasive action because they did not notice a nearby UAV in time, in some cases because they had just passed through clouds while on their final approach to land.

Britain has seen a similar spike in drone incidents. The UK Airprox Board (UKAB), which identifies and analyzes near misses and other cases of potentially dangerous aerial proximity, recorded zero incidents involving UAVs in 2013, six in 2014, 26 in 2015, and 23 in just the first five months of 2016.[108] Of the 49 incidents from January 2015 through May 2016, the UKAB classified 22 as A-level risks, indicating a "serious risk of collision."[109] British airports also reported near misses, such as four at Heathrow, Birmingham and London City in August and September 2015, as drones entered civil aircraft's flight paths.[110] In April 2016, a drone hit the front of a British Airways flight approaching Heathrow, but caused minimal damage and the plane landed safely.[111] Additionally, various local UK police departments reported spikes in drone incidents in 2015, in which UAVs violated British regulations by flying within 150 meters of a populated area or 50 meters of a building,

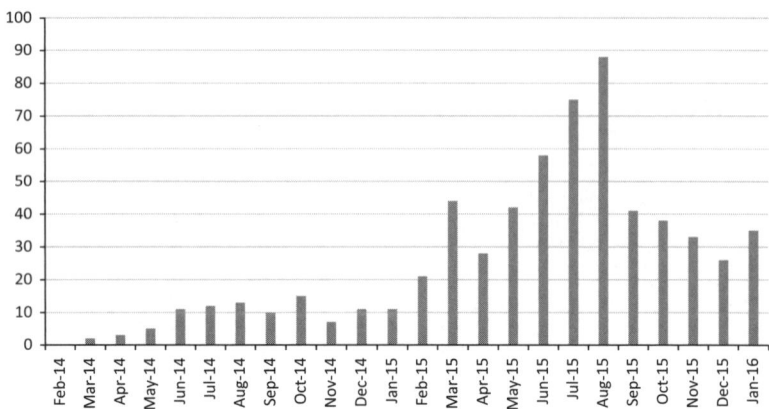

GRAPH 4.2   Airport/Airfield Incidents.

vehicle or individual.[112] Canada,[113] France,[114] and other countries have seen similarly sharp increases in drone activity in restricted airspace. And on October 12, 2017, Canada reported the first drone-airplane collision in North America, about three miles from Jean Lesage International Airport in Quebec City. Fortunately, the twin-propeller plane landed safely, and no one was hurt.[115]

These incidents indicate states are unprepared to prevent terrorists from deliberately flying UAVs into restricted airspace and trying to cause a crash. UAVs in populated areas pose a risk, as a drone could injure people by hitting them—or possibly kill a few if it carried an explosive charge—while drones flying into road traffic could cause car accidents. However, when it comes to small, relatively inexpensive UAVs, perhaps the largest concern involves airports and civilian aircraft.

Fortunately, as of early 2017, no drones have seriously damaged a plane or helicopter in a midair collision. However, civilian aircraft have hit birds, leading to engine failure and crashes. Worldwide, bird strikes cause approximately $600 million per year in damage to planes. According to FAA data, between 2000 and 2009 almost 500 planes collided with birds in United States airspace, and 166 had to make emergency landings. The worst incident in the United States occurred in 1960, when birds got sucked into the engines of Eastern Air Lines Flight 375 as it took off from Boston's Logan International Airport. The aircraft crashed into Boston Harbor, killing 62.[116] Most incidents involving birds cause little damage, and very few cause any

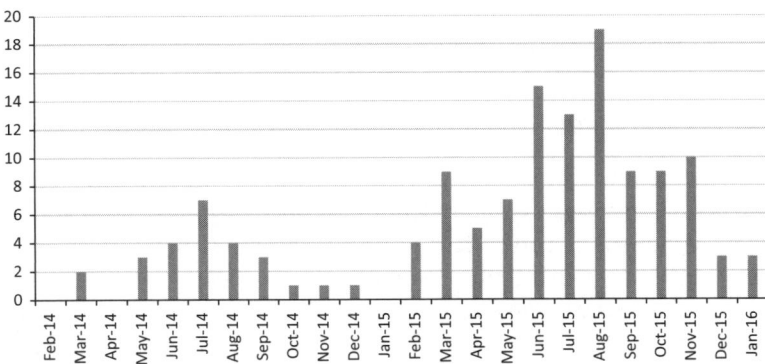

GRAPH 4.3 Close Calls.

deaths. It is therefore likely a midair collision between a plane and a small drone would not be a serious problem, but there is a risk a UAV could crash manned aircraft by hitting rotaries, propellers, particular parts of wings, or by flying into a jet engine.

Terrorist networks, especially Al Qaeda, appear fixated on civilian aircraft. Passenger jets are highly visible targets, symbolizing modernity, freedom of movement, and internationalism; and destroying one garners a lot of attention. They carry many people, sometimes a few hundred, who could all die if an attack destroys the plane. Additionally, if terrorists could force a plane to crash into—or explode over—a populated area, the attack could harm people below along with the passengers. Incidents in which Al Qaeda, ISIS and jihadist sympathizers targeted civilian aircraft include September 11th, AQAP's thwarted 2010 attempt to use bombs disguised as printer cartridges, the November 2015 crash of a Russian passenger jet in Egypt that killed 224,[117] the shoe bomber, the underwear bomber, and more.

This suggests the possibility terrorists will try to fly a small drone into the flightpath of a passenger jet as it takes off or lands, perhaps aiming to fly into an engine. It would not be easy. Commercially available UAVs fly much slower than jets, and pilots can take evasive action, so a successful attack would require precision and luck. However, given the relatively low price of quadcopters and other small drones—and the ability of operators to get away or, if caught, claim it was an accident—attempting attacks would be cost effective and low risk. Airports' inability to prevent UAVs from flying into their airspace demonstrates governments are unsuitably prepared to prevent deliberate attempts to use drones to damage or destroy passenger jets and other civilian aircraft.

COUNTERING THE THREAT OF SMALL UAVS

Clearly, drone regulations are, on their own, insufficient. Legal restrictions provide useful guidelines for operators, law enforcement agencies, air traffic controllers, businesses and individuals. However, laws outlawing drone flights near airports and heliports, in the path of manned aircraft, and over populated areas have not stopped all of those things from happening with increasing frequently. Without sufficient countermeasures and enforcement, legal restrictions cannot provide security. This is especially true in asymmetric warfare, as states face opponents that do not adhere to their laws.

In December 2015, in response to reports of small, privately-operated UAVs violating restricted airspace, the FAA announced the creation of a drone registry. Anyone who owns a small drone (weighing between 0.55 and 55lbs), has to register it using an online system. This registry applies to recreational unmanned aircraft; the FAA plans to set up a separate registry for commercial drones in the future.[118]

These registries will help the FAA regulate private drone flight, but they provide limited value as an anti-terrorism defense. Non-compliance could result in a fine, though whether responsibility for enforcement lies with local, state, or federal agencies remains unclear.[119] The registry of recreational UAVs creates an incentive for operators to avoid accidents, gives police a resource to track down the owner of a drone that crashes in a restricted area, and creates the means to hold operators liable if their UAV causes harm or property damage. However, it is unlikely anyone who plans to use a drone in a terrorist attack will register it in advance, and there are enough ways to buy small UAVs online or from overseas sellers, or build them using instructions from DIY Drones or other websites, that terrorists could skirt registration requirements.

Stopping terrorists from using relatively inexpensive multicopters to gather information would be more difficult than preventing surveillance by people on the ground. Visible signs forbidding photography, along with the presence of police or soldiers authorized to confiscate cameras, represent a deterrent for terrorists trying to case a target. However, while humans may fear getting caught, the same could not be said of UAVs. In cities, it would be difficult for officers to follow drones that can fly over buildings and move horizontally at over 20 miles per hour, and radar would be unable to track them given their size and the obstructions common to urban environments. Security officers could try to shoot them down, but this might be dangerous in a populated area. In the event security services were able to down a small, makeshift surveillance drone, they still would have difficulty finding its operator.

Protecting against the threat of small UAVs thus requires measures to prevent drones from entering restricted airspace and to stop them if they do. Ideas include UAVs designed to capture or disable small drones; firing a net from the ground or from a drone in the air;[120] shooting with guns, perhaps with specially designed shotgun shells;[121] C-RAM systems; and anti-drone laser cannons.[122] The Dutch National Police even started training eagles to snatch them out of the sky.[123] Many of these measures are expensive and/or potentially dangerous, and therefore better suited to defending military bases than civilian areas such as an airport or stadium.

Electronic measures provide a possible alternative. To remotely control an average recreational UAV, or the replica planes Rezwan Ferdaus planned to use in his attack, the operator sends a radio signal, which can be traced using the sort of electronic sensors included on advanced military ISR drones. While the United States controls UAVs from far away with satellite signals, or programs unmanned aircraft to autonomously carry out predetermined missions, less sophisticated drone pilots typically use remote controls with limited range. Pinpointing the signal would enable a response, such as a police car or a missile strike. However, such a technique is rendered moot by ArduPilot and other commercially available autopilots that enable pre-programmed flights.

To stop terrorists and insurgents from using UAVs as kamikaze weapons or ISR platforms, security services first need to identify a drone as suspicious. This would be relatively easier against an insurgency in an active theater of war, in which all authorized aircraft are operated by the counterinsurgents and allied forces. It becomes much more problematic in an environment with legal commercial drone flights, in which the vast majority of UAVs are innocuous, but any one of them could be a flying bomb.

For organizations that deploy suicide bombers, drones present an alternative method of guiding explosives to targets, and would not expend human operatives. A robot in a populated area stands out more than a person, limiting its ability to surprise or to move to a location where an explosion would cause the greatest damage. However, this limitation will decrease as drones become increasingly normal sights, especially in cities. If delivery drones and others undertaking daily tasks become widespread, then the presence of small unmanned aircraft will be commonplace and will not raise any alarms. Furthermore, delivery UAVs would follow automated flight plans, which eliminates the possibility police or military could locate the drone operator by tracing the radio signal used for remote-control aircraft. These, in turn, increase the probability a terrorist will acquire a commercially available drone, or construct one from parts by following online directions, and transform it into a weapon.

While it is difficult to predict exactly who will attempt this sort of attack, it is easier to identify the risk and create countermeasures in advance. In anticipation, states should develop methods of taking control of any robot within a given area in an emergency. Much as the FAA can order commercial aircraft to remain on the ground or human pilots to change course, governments should be able to order unmanned aircraft to land or alter their flight plan as needed. Additionally, as commercial drones proliferate, infrastructure, government buildings, and other potential targets could be outfitted with electronic measures that force any UAV that gets too close to turn around, perhaps by triggering the automated return-to-base function, which many drones already include in case their remote control signal gets cut off or preprogrammed flight plans go awry.

Triggering the return-to-base function, taking over a drone remotely, and other electronic defenses would act like the physical barriers that defend against car bombs by preventing vehicles from getting too close to potential targets. By anticipating the ways terrorists might use commercial robotics technology and developing countermeasures in advance, states can ensure restricted airspace, such as above an airport, remains drone free, while mitigating the risk of a deliberate attack. Additionally, reliable electronic defenses could provide security without instituting onerous regulation that hinders businesses' ability to creatively employ commercially available UAVs.

CHAPTER 5

The Smart SWARM Strategy

Drone strikes make headlines, but the primary advantage of military robotics is informational, not kinetic. Whether operating in the air, on the ground, or at sea, unmanned systems fire the same weapons as their manned counterparts. For example, UAVs strike targets with missiles that could be mounted on fighter jets. Armed drones are useful, because they can attack a target without placing a pilot at risk, but they do not offer substantially different kinetic capabilities, and therefore fit within existing military strategy.

However, unmanned technology can gather, and computers can process, far more information, from far more sources at once, than human beings. Therefore, existing and forthcoming technology has the potential to provide warfighters with a comprehensive informational advantage; a digital map of a given battlespace that automatically updates in real time. An automated swarm of miniature drones—essentially a group of flying cameras, microphones and other sensors—could fly low to the ground, under cover, inside buildings, and close enough to see people's faces, dramatically improving situational awareness. With this complement to larger information-gathering platforms higher in the air, the United States could create a multilayered surveillance pyramid, achieving "information superiority" as sought by advocates of Network-Centric Warfare (NCW). In an asymmetric context, these capabilities would aid counter-insurgents by reducing their informational disadvantage, thereby countering one of insurgents' primary assets.

This chapter lays out the basis of an information-focused strategy for robotic warfare that adapts the principle of mass to informational capabilities. Whereas traditional military strategy focuses on massing combat power at a particular place and time to overwhelm the enemy and hedge against uncertainties, the "Smart SWARM strategy" focuses on massing intelligence, surveillance and reconnaissance (ISR) capabilities to reduce uncertainty and provide the awareness to defeat the enemy with maximum efficiency. Essentially: know, then act; with automated systems providing the identity and location of all people and relevant objects, improving the speed and accuracy of decision making at all levels, from senior commanders to individual soldiers. Though the moral, political, and strategic dimensions of warfare bear little resemblance to video games, this is analogous to the United States using a cheat code that reveals the location, identity, and capabilities of all other players, while those players are still operating in the dark.

NETWORK-CENTRIC WARFARE (NCW)

The United States has made information sharing and widespread battlefield awareness a priority since the 1990s under the rubric of "Network-Centric Warfare," which was coined by Arthur Cebrowski and John Garstka, and developed by the Office of Force Transformation under Cebrowski's direction.[1] Drawing upon internet-era theories of business, economics, and sociology, NCW emphasizes using modern information technology to share information and coordinate actions among various units and platforms in real time, so the whole of a military force is greater than the sum of its parts. Ideally, transforming the military to a seamlessly networked force would eliminate the fog of war by removing the friction between separate military units and their commanders. According to Cebrowski, NCW represents no less than "transforming from the Industrial Age to the Information Age," in which "power is increasingly derived from information sharing, information access, and speed, all of which are facilitated by networked forces."[2] By linking and coordinating the military's various parts, Cebrowski and Garstka theorized, a smaller, more mobile fighting force could command the destructive power of larger armies, but with greater speed and precision, granting the United States armed forces a qualitative military advantage.

The core strategic goal of NCW is "information dominance" or "information superiority." According to the United States military's "Joint Doctrine for Information Operations" (known as "Joint Pub 3-13"), this refers to "the ability to collect, process, and disseminate an uninterrupted flow of information while exploiting and/or denying an adversary's ability to do the same."[3] Essentially, the more soldiers, sailors, Marines, airmen, and their commanders know about a given battlespace, the more efficiently they can pursue their objectives. Increasing knowledge reduces the risk of mistakes, such as friendly fire, by informing engaged units of their allies' locations, and allows forces to act with greater speed and precision by reducing ambiguity. Perfect information is a limit, a goal that can be approached but never reached, and information dominance works to serve and enhance traditional war-fighting concepts, rather than replace them. As NCW advocates David Alberts, John Garstka, and Frederick Stein point out, "even in the case where information is far less than perfect, it could reasonably be argued that being able to have a shared understanding of what is known and what is not known would be preferable to a situation in which units operated in isolated ignorance."[4]

Network-Centric Warfare received its first major tests in Iraq in 2003, garnering numerous critics. For example, Brookings scholar P.W. Singer declares networking-based strategies a "failure" and dismisses the "networks of email and Internet fiber optics that now bind military units together" as merely quicker versions of "radios, phones, or faxes."[5] The "network crowd," he argues, "was wrong that the fog of war would be lifted,"[6] citing instances in which American forces lost track of Iraqi tanks in the early conventional phase of the invasion, and then had difficulty identifying and tracking insurgents after the defeat of Saddam Hussein and the Iraqi army. Defense analyst Loren Thompson goes further, arguing that NCW was conceived before 9/11 and designed for conventional warfare, and thus should not have surprised anyone when it performed poorly against asymmetric adversaries like the Iraqi insurgency.[7]

Part of the problem was the Bush administration's, especially Secretary of Defense Donald Rumsfeld's, overzealous conviction that NCW was ready in time for the invasion of Iraq. In Operation Iraqi Freedom, the equipment for network-centric strategies ran into both technical and logistical problems. "Rather than a seamless flow of information, soldiers wrestled with everything from Web browsers constantly crashing due to desert sand to heat fouling up equipment designed for use in offices, not battlefields."[8] Additionally, the newly networked fighting forces faced unexpected bottlenecks, from a shortage of batteries to power mobile devices, to an overwhelmed bandwidth spectrum for wireless communications. The Defense

Department did not fully anticipate these problems because it had not tested NCW in actual fighting conditions; and the war in Iraq revealed ways in which the military fell short of Cebrowski's vision. Though this is a legitimate criticism of the confidence with which some officials championed military transformation in the early 2000s, it presents technical problems to be overcome rather than a reason to discredit the larger theory.

More problematic were difficulties in identifying and tracking adversaries, especially in the post-invasion insurgency phase of the Iraq War. In *Armed Forces Journal*, Milan Vego argues NCW "appears not to provide much of an advantage in fighting an insurgency in the post-hostilities phase of a campaign, as the current situations in Afghanistan and Iraq illustrate. In fact, the ongoing insurgency in Iraq is a powerful proof, if any is needed, of how little practical value networking one's forces has in obtaining accurate, timely and relevant information on the enemy."[9] Casualty data support this, as the United States defeated the Iraqi army and deposed Saddam Hussein in two months, losing 187 Americans, but suffered 4,299 military deaths while fighting the subsequent insurgency for eight and a half years.[10]

THE POTENTIAL BENEFITS OF ROBOTIC TECHNOLOGY

In *Wired for War*, P.W. Singer argues the United States lacks a robotic warfare strategy designed to achieve military goals utilizing the new capabilities robots provide.[11] Analogously, tanks and airplanes appeared in World War I, but only in support of existing attrition and trench-warfare strategies, scouting ahead or supporting infantry and artillery. It was not until World War II that Germany built a strategy around these new technologies, utilizing their strength and speed to target the political and industrial support base of national war efforts. Even though France had more tanks than Germany in World War II—3,245 to 2,574—French doctrine dispersed a few to each infantry unit, while the German blitzkrieg coordinated tanks with planes and artillery "to create a concentrated force that could punch through enemy lines and spread shock and chaos."[12] German forces went around, over, or through French defenses, rapidly taking Paris and conquering all of France in less than two months. Citing soldiers, generals, and roboticists, Singer warns that developing the right strategy "for using unmanned systems is thus essential to the future" of American security, so the United States does not develop "the Maginot Line of the twenty-first century."[13]

Currently, the United States military uses robots to enhance pre-robot strategies. Existing units get small unmanned aerial vehicles or links to larger robotic airplanes for scouting; ground-based bots to carry equipment or scout ahead with cameras; and specialized robots designed to meet current needs, such as searching for and disposing improvised explosive devices (IEDs). Meanwhile, aerial drones have taken over many of the missions previously conducted by piloted aircraft, from long-range reconnaissance to airstrikes. The military thus uses robots to enhance previously developed capabilities and reduce the risk to personnel, but does not yet have an overarching strategy "on how to use them or how they fit together."[14]

In part, this is the result of thinking about how humans would use robots to fight, not how computers would fight a war. The main advantages granted by robotics are reduced risk to human personnel, and increased information gathering and processing. Unmanned systems do not significantly improve destructive capability. Any weapon attached to a land, sea, or air-based robot could be carried by a human soldier or manned vehicle. Bullets, bombs, and missiles, along with potential futuristic weapons like lasers, could be used by humans and robots alike to kill and destroy. There are no counter-measures that can sufficiently protect against these destructive capabilities—many targets are unarmored, and larger or more directed explosives, such as bunker-busters or shaped charges, can destroy those with physical protection—which means destructive capacity would not distinguish computerized warfare from the pre-robot doctrine of the early twenty-first century. Since the invention of precision-guided munitions (AKA "smart bombs"), advanced militaries have been able to quickly destroy a known target at will. The problem is knowing what to target.

Therefore, gathering and processing information should form the basis of robotic warfare strategy. Like human commanders, a computer fighting a war would want as much information as possible, avoiding action without sufficient information when it can, and filling in information gaps with assumptions and educated guesses when it must. However, unlike humans, a computer can simultaneously utilize as many streams of information as software and processing power allow. Taking this idea to its theoretical limit, a network of robots integrated with a powerful computer system could fight with perfect information.

The ideal of fighting with complete information is not new— arguably it has been a goal of militaries since the first organized fighting forces sent out scouts—but technology is finally approaching the point at which something close becomes plausible. Unmanned systems can complement humans' information-gathering efforts to provide a more comprehensive awareness of friendly forces, enemies, and civilians, along with relevant objects such as weapons and obstacles, within a given battlespace.

Imagine a swarm of flying robots: hundreds or thousands of micro UAVs, the size of dragonflies or smaller insects, equipped with sensors, cameras and microphones, feeding information into a supercomputer. These intelligent Systematic Warfighter-Assisting Reconnaissance Measures, or "Smart SWARM," could fly in front of advancing forces and provide detailed, real-time information about what lies ahead. Taking advantage of computers' ability to simultaneously process more streams of information than humans, portions of the swarm could fly off in all directions, gathering information from all sides. The computer could then account for any people and objects within range, and create a fluid three-dimensional display that changes as circumstances evolve. It could present the full 3D image or a two-dimensional bird's-eye view on screens in a command center, and display a simplified version to soldiers in the field in a manner that would not obstruct their vision, such as a hand-held device or a headset display similar to augmented reality glasses. Depending on the needs of various users, the system could utilize algorithms to prioritize the likelihood of threats, alerting soldiers or commanders when something requires their attention. For example, with information-gathering drones surrounding soldiers at periodic intervals, this system could alert units to approaching enemies, track fleeing suspects, and discover awaiting ambushes.

Additionally, the Smart SWARM could enter buildings before soldiers, informing them of the locations of any people or weaponry. With many small, mobile robots, the SWARM could provide a real-time map of the inside of any building, and individual micro UAVs could follow people as they move around inside. Using object recognition software, the system could determine which individuals are carrying weapons, and locate any guns or other relevant objects stored on the premises. With infrared sensors, it could find anyone waiting to ambush soldiers or hiding to escape capture. Add "Fido sniffers,"[15] which explosive ordinance disposal robots currently use to find bombs,[16] and the SWARM could identify booby traps and discover stored explosives. This would provide commanders with detailed information about the contents of any house or compound before sending in soldiers, greatly reducing enemies' ability to surprise or flee, and increasing the thoroughness of searches.

In addition to increasing the efficiency of kinetic operations, the SWARM would aid intelligence operations and facilitate new spying techniques. A system of small, flying robotic sensors, cameras and microphones could eavesdrop on various targets, observing meetings or hiding in suspected safe houses. They could sneak through pipes or between walls, resembling small insects to reduce suspicion. This would provide a method of bugging targets without the risk a human operative could be caught planting or recovering the recording device. Furthermore, the bug would not be fixed to a single location, and a human operator or computer program could direct it to move as the target moves, or if the audio or video feeds become obstructed. The system could thus monitor terrorist or insurgent leaders targeted for drone strikes, or snatch-and-grab missions, confirming the identity of the target and informing decision makers of anyone in proximity, thereby reducing mistakes and limiting collateral damage.

It would be naïve to assume the Smart SWARM could eliminate the fog of war, but it could reduce it considerably and provide relative information superiority. With robots gathering information and computers processing the numerous streams, the United States military could achieve a considerably higher level of battlefield awareness than in the Iraq War, in both the first phase against the Iraqi military and the second phase against the insurgency. This would facilitate more efficient operations, achieving goals with fewer military and civilian casualties.

Nevertheless, pre-robot intelligence techniques would remain essential to strategies against insurgent and terrorist organizations. A widespread network of information-gathering robots could monitor the location of people and objects, and a sufficiently intelligent computer program could identify actions such as carrying weapons or planting IEDs. However, robots could not determine intentions or allegiances. Therefore, human intelligence techniques, from infiltrating hostile organizations to establishing relationships with locals, would remain essential to counterinsurgent strategy.

Information gathered in traditional ways would help direct, and be informed by information gathered and processed by robotic systems.

Of course, perfect information is impossible. To take an extreme example, mind reading would greatly enhance military tactics by revealing opponents' intentions, but acquiring this information is far beyond the scope of any existing or forthcoming technology. However, with larger intelligence, surveillance and reconnaissance systems high in the air and hundreds or even thousands of information-gathering robots on or near the ground, each carrying a variety cameras and sensors, all linked to a powerful information-processing computer, a fully roboticized military could achieve real-time awareness of people and objects within a given area. This would allow more informed decision making, on a larger scale, than any military in the history of the world.

ROBOTIC CAPABILITIES AND INTEGRATED COIN STRATEGY

Hundreds of miniature flying robots working in conjunction to create a 3D map of all people and objects in a given area that updates in real time may sound like science fiction, but most of the elements already exist. Micro UAVs designed to gather information already fly in active military theaters, such as the Black Hornet miniature helicopters British soldiers operated in Afghanistan.[17] Prototypes of UAVs that look like and mimic the abilities of various insects have been built, and more are expected soon. Groups of small robots independently working together to accomplish a single task already exist, though they have not yet been mass produced or put to widespread use. Object and facial recognition software already exists, and is advancing rapidly. Finally, though it appears no software capable of constructing constantly updating 3D maps from all of these inputs currently exists, various existing programs suggest such a thing is possible.

Micro UAVs

In addition to more expensive models, such as the Black Hornet, micro UAVs have begun proliferating. A team from Georgia Tech's Robotics and Intelligent Machines Department has developed a miniature UAV that mimics the flying capabilities of a dragonfly. Developed with a $1 million grant from the US Air Force's Office of Scientific Research, the Dragonfly drone can fly and hover in a manner similar to its namesake.[18] It employs a combination of helicopter, multicopter, and fixed-wing technology to achieve considerable maneuverability, and, with a length of six inches and a weight of 25 grams, is small enough to fit on a human palm.[19] TechJet, a company spun off from the Robotics and Intelligent Machines Department to market the Dragonfly, envisions various versions tailored to different uses, including gaming, photography, home security, and military surveillance.[20]

With its small size, ability to both fly and hover, and flexibility regarding components, the Dragonfly, or an alternative UAV with similar capabilities, could form the basis of the Smart SWARM. TechJet expects the Dragonfly to retail between $250 and $1,500, depending on the level of computing and flying capabilities included.[21] This makes the high-end version far less costly than the Black Hornet, which the UK bought for approximately $200,000 per unit. Unlike the Dragonfly, each Black Hornet includes a camera that can capture video or still images from a kilometer away and a hand-held display that can show these images to an operator. Although the Dragonfly does not come with these accessories, its considerably lower price demonstrates forthcoming UAVs based on insects could provide a cost-effective basis for building the Smart SWARM. Furthermore, since the distance between drones in the SWARM would be much less than one kilometer, they would not require the Black Hornet's powerful camera.

An even smaller alternative is a tiny flying drone called RoboBee created by the Wyss Institute for Biologically Inspired Engineering at Harvard. This micro UAV is only three centimeters from wingtip to wingtip and weighs only 80 milligrams. Whereas the Dragonfly is a little smaller than a human palm, RoboBees are barely larger than a penny. Unlike miniature helicopters and other micro UAVs that use rotaries, it is an ornithopter, with flapping wings.[22] The RoboBee flies using "'artificial muscles' that contract when a voltage is applied," and the Wyss Institute is working towards more complex structures modeled after insects' wings, including flexible veins and membranes.[23] This miniature drone can already hover and execute simple flight maneuvers, demonstrating the feasibility of flying robots tiny enough to escape casual notice. Though the Dragonfly or RoboBee would not be confused with actual insects by anyone paying attention, as UAVs get smaller and closer in appearance to real insects, they will become increasingly capable of stealthy spy missions.

As UAVs' computer chips become more sophisticated, drones will become increasingly capable of independently executing complex tasks, such as navigating an unknown space or following a moving target. Many commercially available UAVs already include chips with autopilot capable of executing a pre-programmed flight plan or autonomously returning to base using GPS if the drone loses communication with its remote pilot.[24] More sophisticated UAVs, such as the US Air Force's RQ-4 Global Hawk, can autonomously survey thousands of square miles in a single flight.[25] New "neuromorphic" chips, designed based on features of the brain, enable smarter drones, capable of greater environmental awareness and learning.

In November 2014, HRL Laboratories' Center for Neural and Emergent Systems announced a successful test of a small drone equipped with a neuromorphic chip.[26] Upon entering a new room, a micro UAV utilized HRL's chip to process data from optical, ultrasound and infrared sensors to autonomously determine it was in a new location. The new data in this particular combination caused changes in the chips' "neural" architecture, which meant the second time the drone entered the room it recognized the location. The chip used only 50 milliwatts of electricity—much less than a laptop computer would require to run software capable of recognizing a specific room.[27] The success of this test demonstrates the feasibility of micro UAVs smart enough to populate a Smart SWARM.

Robotic Swarms

To fulfill the Smart SWARM's goal of information superiority, a sizeable group of these small robots would need to saturate a given battlespace, autonomously coordinating their actions. This is simpler than it might seem at first. To act as a swarm, robots do not require constant instructions from a centralized decision maker. In a manner similar to the way bees or ants work together, each robot follows some simple rules in relation to other units in the swarm, which, when taken together, produce collective action.[28] For example, ants carrying food back to the main colony just follow the ant in front of them along a chemical trail left by the ants that originally discovered the food source and reinforced by every ant that walks along the path. Similarly, each robot in a swarm of information-gathering micro UAVs could be programmed with a simple rule: never get too close or too far from another member of the SWARM. This would keep them close enough to cover a selected area with no blind spots, but spaced out enough to avoid crashes or unnecessary redundancy.

There has been considerable research into robot swarms, much of it published in robotics journals, including some specifically devoted to swarming technology. In particular, the Future and Emerging Technologies program of the European Commission sponsored a venture called "Swarm-bots," and a successor called "Swarmanoid," to advance coordinated robot behavior. The Swarm-bots project focused on homogenous groups of robots that autonomously assembled themselves into a single structure.[29] Swarmanoid built upon this by creating a heterogeneous swarm consisting of approximately 60 robots of three different types that moved as a group through human environments while working together to negotiate obstacles.[30] The Swarmanoid system won an award from the Conference on Artificial Intelligence in 2011, and its success indicates the feasibility of an autonomous swarm featuring different types of information-gathering micro UAVs.

Drawing lessons from the behavior of insect societies, programmers have developed groups of robots that can make decisions from a collective process of individual actions. For example, researchers working on the Swarm-bots project noted how group decisions emerge from the interactions of individual ants seeking the most efficient path to a food source. No single ant knows the best way to reach the food, but, through trial and error and the ability of individuals to chemically communicate success or failure, the group of ants finds and then adheres to an effective route.

With this in mind, researchers programmed a group of ground-based robots with simple rules that enabled the group to avoid falling into holes. As each robot moved across terrain pockmarked with holes from which they could not escape, some inevitably fell in. However, any that fell into a hole would signal to other robots to keep their distance, which resulted in most of the group avoiding the holes and reaching their destination. The group's decision to steer clear of the holes emerged from the simplistic actions of, and signals from, individual members.[31] In another example of emergent group decision making, a heterogeneous swarm improved group efficiency by increasing the role of machines that proved the best at performing various tasks, independently creating a division of labor.[32] Building upon this, recent developments in "modular robotics" have created swarms in which small robots act as reconfigurable "building blocks" that autonomously assemble themselves into larger, more complex structures.[33]

The ability to learn from circumstances and reassign roles based on performance would make a robotic SWARM capable of reacting to combat conditions. Some early ideas for swarms depended on a centralized decision maker that would fly above the other robots to provide information about the surrounding environment and issue directions. However, if one these central robots sustained damage or malfunctioned, the entire swarm would suffer. More recent robot swarms based on emergent group decision making are more robust, which makes them better suited for military operations. The entire swarm is valuable, but each individual unit is incidental. If a few became damaged, whether through accident or hostile action, the group could adjust and continue with its mission.

Moving a swarm through most outdoor areas is relatively easy compared to navigating indoor environments. However, a paper presented at the 2012 International Conference on Robotics and Automation detailed an "entirely decentralized approach" of moving around inside that "relies solely on local sensing without requiring absolute positioning, environment maps, powerful computation or long range communication."[34] The authors reported successfully testing this method using quadcopters. In the experiment, the drones in the swarm did not possess any prior knowledge of the halls they moved through, but were able to navigate the indoor space using basic information each robot gathered about its environment and simple rules governing relations between the units of the swarm. Using an advanced version of this application, a Smart SWARM would be able to enter a building and move throughout it, searching for people and objects while providing a layout to human operators.

Micro UAVs like the Black Hornet and Dragonfly are limited by small batteries, which typically enable sustained flight for only 20–30 minutes. Though the technology is advancing, a lightweight battery that could provide enough energy to keep a small UAV aloft for many hours does not appear likely in the near future. Therefore, to maintain continuous, comprehensive surveillance of a given battle-space, the Smart SWARM would need to swap in fresh drones for exhausted ones. Most commercially available UAVs, such as small quadcopters designed for aerial photography, already come equipped with an automatic return-to-base function for when their battery level drops or if they lose contact with their operator. With a ground station, land vehicle or flying "mothership" housing backup units, the micro drones that make up the SWARM could cycle between surveillance and recharging, ensuring uninterrupted coverage of a desired area.

Though this has not been utilized on a scale sufficient for comprehensive battlespace surveillance, a mobile base for small UAVs already exists. In 2015, the Polish company WB Electronics announced it had designed a system called "Bee," in which multiple small drones can launch from a military vehicle, such as an armored personnel carrier. All drones in the Bee system carry cameras and provide surveillance, and some carry an explosive charge as well, capable of detonating midair on command. The ground vehicle thus provides a method of transferring small UAVs with limited battery life close to a target. With staggered launches in which spent UAVs return to base and charge or swap out their batteries, the Bee could provide a means by which the Smart SWARM maintains continuous coverage. This technological limitation will presumably decrease over time as battery technology continues to develop.

Additionally, the Bee includes an independent communications system, which enables it to operate in a variety of environments. According to Wojciech Komorniczak, WB Electronics' Director for Research and New Technologies, "battle conditions are, for instance, when there is no mobile phone connection and the generally available wireless network is gone, so we have to supply such a network by ourselves."[35] An independent communications system helps address one of the technological shortcomings that hindered Network-Centric Warfare in the invasion of Iraq.

For more rapid deployment, the SWARM could utilize a system tested by the US Navy in 2015 known as Low-Cost UAV Swarming Technology, or LOCUST, which launches up to 30 small drones out of tubes similar to those that fire missiles. According to the Office of Naval Research, these launch tubes could be mounted on ships, ground vehicles, and manned or unmanned aircraft.[36]

Another alternative is a US Navy system built around the Perdix, a propeller plane about half a foot long, with a foot-long wingspan, weighing 290 grams. The Perdix drone can fly up to 70 mph on its own, but can withstand speeds up to 460 mph, and temperatures as low as -10°C, allowing manned planes to launch them midflight. In an October 2016 experiment, three F/A-18 fighter jets deployed 103 Perdix drones, which then executed swarming maneuvers.[37]

Additionally, a vehicle that deploys smaller drones could provide a communications intermediary large enough to transmit information gathered by the Smart SWARM to a distant command center. One possibility is an airship floating at higher altitudes. In 2016, Amazon received a patent for an "airborne fulfillment center," effectively a warehouse blimp floating at 45,000 feet, which would deploy delivery drones.[38] Though Amazon has not built one yet, the idea presages an airship that could transport a Smart SWARM while also serving as a communications hub, processing information from drones below and providing it to ground forces, or transmitting it to satellites above.

3D Mapping

With robotic swarms capable of independently navigating both outdoor and indoor environments, the next step towards a militarily useful Smart SWARM is software that synthesizes the various streams of information to create a user-friendly display. Many of the components for this exist as well. Architects, land developers, city planners, miners, and other professions that make use of geographic and spatial data commonly use three-dimensional mapping software. The maps this software creates are static models of topographical features, city blocks, or buildings, rather than the constantly updating displays needed to monitor dynamic environments.[39] However, existing software can build detailed 3D maps users can navigate virtually, and a more advanced version could use this template while adding the ability to track moving objects in real-time.

3D mapping software is primarily a tool for human users to build virtual models, but computers have demonstrated the ability to autonomously create 3D displays of indoor spaces using limited information. In a May 2013 paper published by the National Academy of Sciences, a team presented an algorithm that "reconstructs the full 3D geometry" of a room "from a single sound emission." Modeled after bats' echolocation, the software records sounds bouncing off walls and uses the information to build a three-dimensional map of a room.[40] With all the noise present in a combat environment, this technique might not be the best fit for warfighting. However, this software shows how a computer system can quickly create a model of a room using information acquired by automated processes, and the Smart SWARM would be able to use information

FIGURE 5.1 A Perdix drone, one of many in an experimental drone swarm

from various sensors and cameras, instead of just a few microphones monitoring the echoes from a single sound.

To build three-dimensional images using the information collected by various robots in the Smart SWARM, programmers could adapt software used to study hurricanes. In 2013, a team from the University of Florida announced it is working on a project to predict the strength and path of powerful storms using a swarm of robots. Combining micro UAVs that fly into the storm with small submersible robots that swim in the ocean below, the hurricane-hunting swarm uses sensors carried by each robot to collect data on air pressure, temperature, humidity, wind speed, and wind direction. As with other robotic swarms, each unit is relatively cheap: only $250 per miniature plane. If any are lost—which is to be expected when flying in extreme weather—the group adjusts autonomously. The swarm sends the data it collects in real time to computers out of the storm's range, which then create sophisticated weather models predicting the hurricane's intensity and trajectory.[41]

This system has not been completed yet, but an existing method of studying hurricanes demonstrates how a computer can build a model of complex phenomena using data collected by many small units. To gather data on powerful storms, a large manned airplane flies into the eye and ejects hundreds of lightweight cylinders known as "dropsondes." The dropsondes fall through the storm attached to parachutes, going wherever the winds blow them, and gather data they send back to base via radio signals, whereupon a computer uses this information to create a model of the hurricane.[42] This offers a potential model for how the Smart SWARM's central computer could employ real-time data coming in from multiple sources to create a detailed map of a given area.

Object and Facial Recognition

To provide warfighters with information superiority, the Smart SWARM would need to do more than create 3D maps of indoor and outdoor spaces. These maps would be useful on their own, but would require many human analysts to monitor the images, failing to take full advantage of computers' information processing advantages and undermining the strategic goal of maximizing efficiency. However, with object recognition software the system could autonomously identify people, weapons, and other items of interest, bringing them to the attention of soldiers and commanders, while facial recognition could help the system identify suspects ahead of airstrikes or ground raids. As with the other components of the Smart SWARM, promising versions of recognition software already exist.

In the twenty-first century, there have been impressive developments in the field of computer vision, which seeks to teach machines to replicate humans' ability to understand components of visual imagery. While humans have little difficulty distinguishing foreground objects from the background in pictures or videos, computers require complex algorithms to accomplish this basic task. It is especially challenging to teach a computer to recognize the same object from multiple angles, at multiple scales, or when partially obscured.[43]

Before long, computers will be able to identify objects without human direction. In 2012, Google received a patent for software that autonomously identifies objects in videos on YouTube, its video-sharing site. Instead of asking users to label objects in their videos, the software utilizes a database of "feature vectors," such as color, shape, texture, and movement, to compare various objects across videos and label them automatically.[44]

A complementary project at Google X, Google's forward-looking research lab, designed artificial intelligence software that taught itself to recognize cats in photographs. Using 16,000 connected processors, the "neural network" software scanned 10 million randomly selected digital images and, over time, figured out how to identify cats in a manner similar to how the brain learns to understand images. According to Google fellow Jeff Dean, "we never told it during the training, 'This is a cat.' It basically invented the concept of a cat," gradually assembling the idea of what a cat looks like by "employing a hierarchy of memory locations to successively cull out general features after being exposed to millions of images."[45]

These advancements demonstrate how computer vision could provide an advantage to advanced militaries in asymmetric warfare. The Smart SWARM's many simultaneous streams of information could overwhelm human analysts, who may be unable to process all the data, determine what is important, and provide engaged soldiers with relevant information in time for them to act. However, a sufficiently powerful computer could. Like cats, guns are distinct objects that come in many varieties but share essential characteristics. A computer system that could learn to identify all firearms within a given area, such as the inside of a building, could process the images gathered by a swarm of drones, alerting soldiers to the guns' locations—perhaps by highlighting them in a bright color in a headset display—and determining whether they are in someone's hands.

Subsequent developments in computer vision include the ability to identify entire scenes, such as a group of men playing Frisbee or a herd of elephants walking across a field.[46] Software that can pick out individual objects from the background is useful, but falls short of human vision, which can place the objects within a larger context. Announced in 2014 by a team from Google and Stanford University, scene recognition software, using neural network processing like the software that taught itself to identify cats, created captions for pictures, many of which were similar to captions created by humans. There is considerable room for improvement—for example, while the software correctly identified an Ultimate Frisbee game, it oddly labeled a kite with a face painted on it as a man flying on a snowboard—but the rapid developments presage software capable of identifying actions, such as an individual planting a bomb or a group aiming weapons.

The next step requires improvements in what is known as Explainable Artificial Intelligence (XAI). Neural networks give computers the ability to teach themselves, but they are not yet able to explain their thought process to humans.[47] For example, Google's neural net learned to recognize cats, but could not articulate its rationale (the object in the image has four legs, fur, claws, its facial structure differs from a dog's, etc.) Without XAI, the military value of video-monitoring software is limited. A neural net might learn to recognize what it believes are suspicious vehicles, but if it cannot explain its assumptions and logic to human operators, military commanders cannot rely on it for life-and-death decisions, nor sufficiently evaluate its performance. Recognizing XAI's importance, an April 2017 memorandum from the United States Deputy Secretary of Defense established the Algorithmic Warfare Cross-Functional Team (AKA Project Maven) to advance machine learning, especially regarding video analysis.[48]

Beyond recognizing objects and scenes, computers are getting better at identifying individual faces. With software that scans facial photographs, law enforcement services, including the FBI, have used passport imagery to catch fugitives in foreign countries. Using a technique known as Elastic Bunch Graph Matching, software can now piece together a three-dimensional facial composite based on partial photographs from various angles, such as those captured by closed circuit television cameras.[49] In another example, Facebook designed a "facial verification" program known as "DeepFace." The software autonomously identifies two images of the same face, and performs nearly as well as humans. While people identified two unfamiliar photos of the same person 97.53% of the time, DeepFace succeeded in 97.25% of cases.[50]

Combined with a Smart SWARM, facial recognition software could prove especially useful for targeted killings or capture missions. The system could identify subjects, letting ground troops know if their target is present before initiating their mission. Additionally, after identifying an individual, a system with object recognition software capable of understanding scenes could follow the target, tracking his movements until he could be apprehended. Such a system would provide commanders with the ability to make more informed decisions, minimizing collateral damage and risks to friendly forces.

These developments in computing and robotic systems demonstrate the Smart SWARM is technologically plausible. Many types of small UAVs that gather information, swarms of robots that coordinate the actions of different types of autonomous machines without central direction, computer software that creates three-dimensional maps of indoor and outdoor spaces, object and facial recognition software, and algorithms that create models from information gathered from numerous sources all exist. These components could be combined and further developed to produce a robotic swarm and dedicated computer system designed to provide soldiers and commanders with a detailed picture of a selected area. Such a system could help the United States military achieve information dominance in both symmetric and asymmetric conflicts.

INFORMATION-FOCUSED ROBOTIC WARFARE AND COUNTERINSURGENCY

In the article "Air Force Strategic Vision for 2020–2030," Air Force General John A. Shaud and Air Force Research Institute professor Adam B. Lowther identify "global situational awareness" as one of the "five critical capabilities" the United States should focus on over the next two decades.[51] The goal of global situational awareness is "the understanding of the strategic, operational, and tactical environments gained through the use of space, air, sea, land and cyber information collection systems."[52] Similarly, the Defense Department's "Unmanned Systems Integration Roadmap FY2013-2038" recognizes "a metamorphosis is needed to develop into a tightly organized and dynamic ISR force."[53] In particular, the Integrated Roadmap calls for development of advanced computer systems that can process the flood of incoming information and "manage this glut of data," arguing it "should become an Air Force funding priority."[54] These documents indicate the United States military understands the importance of information gathering and processing to twenty-first-century warfare.

This goal is worthwhile on its own, but will be especially useful if the United States incorporates improvements in information gathering and processing into a larger strategy based on information dominance. Technology is reaching the point at which an information-focused, robot-enabled version of Network-Centric Warfare can reach its full potential. NCW aided American military operations in Iraq and Afghanistan by coordinating the activity of various military units and platforms, but did not fulfill its advocates' promises. Information moving through the networks improved situational awareness, but the benefits were limited by the inputs.

However, a Smart SWARM, in conjunction with a variety of larger robotic ISR platforms, could create a multilayered surveillance pyramid that can gather and process enough information, at fast enough speeds, to facilitate an information-focused strategy. Know, then act; with satellites, upper atmosphere platforms, and larger ISR drones providing general awareness, and swarms of smaller UAVs supplying more detailed information closer to the ground. By massing these informational capabilities at points of interest, the United States could acquire much greater situational awareness than its opponents, allowing American forces to act faster and more decisively, detect and respond to enemy actions quickly and effectively, accomplish various goals with less risk to personnel, and reduce error.

A Multilayered Surveillance Pyramid

The asymmetric wars in Afghanistan and Iraq demonstrate network-centric operations did not overcome insurgents' informational advantages, but that is because the information fed into the network was gathered primarily by humans. Though valuable, it proved insufficient. As defense analyst Loren Thompson put it, "all those networks the Pentagon was planning are just conduits" and "what matters more for victory is the accuracy and completeness of the information moving through the networks."[55] Information gathered by human soldiers, and by humans watching video feeds from cameras mounted on soldiers' helmets, satellites, or spy planes, is inherently incomplete and of limited accuracy. When directing combat, senior officers can follow the real-time location of friendly forces on interactive maps (not unlike those of Google Earth), employing GPS to monitor the location of personnel and equipment. This facilitates greater coordination of friendly forces, but the absence of data on the location of enemy forces limits its utility. Uncertainty regarding the location of enemy forces creates problems even when the identity of the enemy is known; it is especially problematic in counterinsurgent warfare when there is difficulty distinguishing enemy fighters from civilians. As long as the identity and location of enemy forces is unknown, perfect networking does not get the United States military remotely close to the ideal of information dominance.

The Smart SWARM could thus provide the missing link in a multilayered surveillance pyramid: satellites on top providing geospatial intelligence and relaying information; a few larger drones equipped with a wide area airborne surveillance system and a variety of sensors offering a view of the ground from 20,000 feet; and some

Smart SWARMs close to the ground to provide comprehensive information about a target requiring further investigation. With all of this information processed by a dedicated computer system—monitored and directed by intelligence and command personnel—the United States could acquire information of high enough quality to track enemy movements and to avoid the sort of intelligence failures that lead to accidental targeting of civilians or friendly forces.

Larger ISR drones typically operate at higher altitudes, using a variety of cameras and sensors to gather information about activities on the ground. Employing wide area airborne surveillance systems, such as Gorgon Stare or the ARGUS-IS camera, they can monitor areas as wide as 100 square kilometers. At such heights, they can miss important details, especially when obstructed by roofs, trees, awnings and other cover.

The Smart SWARM can fill in the blanks left by these larger ISR systems, reducing the uncertainties that cause hesitation, inefficiency, and mistakes. Combined with sophisticated information processing, this combination could provide battlefield commanders, intelligence analysts, and individual troops with a detailed portrait of a designated area, helping to address some of the shortcomings of American counterinsurgency and counterterrorist activity in Iraq and Afghanistan. With larger ISR systems providing a bird's-eye view, and the Smart SWARM massing informational capabilities at points of interest, the United States military could establish information dominance over a building, city, or an entire theater of operations.

Combining a Smart SWARM with a wide area airborne surveillance system could thus provide considerable advantages in the kinetic aspects of counterinsurgent warfare. While larger ISR drones can cover a wide area, they look down from thousands of feet, limiting what they can see, especially in urban settings. But the SWARM could peer around corners, fly over rooftops, search under awnings, look in windows, and enter buildings, providing information on the location of snipers or hidden enemies.

Where American, British and allied forces in Iraq and Afghanistan used one or two small drones to scout ahead, ground forces advancing at the center of a Smart SWARM would be surrounded by information-gathering systems. Drones flying in all directions, searching for weapons and explosives, and distinguishing armed and unarmed individuals, would give counterinsurgent soldiers advance notice of ambushes, and help them avoid harming civilians.

This would shrink insurgents' informational advantage, undermining their ability to successfully attack counterinsurgent forces.

The potential drawback is losing an element of surprise. Though micro UAVs will probably get smaller, faster and stealthier over time, a Smart SWARM would be fairly conspicuous. Its presence could tip off enemies that they are being watched, and that an attack could follow. The SWARM partially compensates for this by providing such comprehensive information that enemies would have difficulty fleeing in advance and would remain at a disadvantage once the engagement began; and targets would still not know the nature of any forthcoming attack. However, commanders would have to weigh the SWARM's informational advantages against the disadvantages that come from alerting enemies to its presence, perhaps refraining from deploying the SWARM when they deem the intelligence provided by less conspicuous ISR measures, such as a larger drone higher in the air, to be sufficient.

Reducing Error

Mistaken or misinterpreted aerial surveillance has played a role in some of the United States military's most egregious errors. Among the most infamous was an attack in February 2010 that killed 23 Afghan civilians—members of a religious minority who were fleeing the Taliban in a small convoy. This incident illustrates the danger of acting on insufficient information, using assumptions to make up the difference.

In the early morning of February 21, 2010, as a Special Operations team was approaching the village of Khod in the Afghan province of Uruzgan, military intelligence picked up a broadcast radio message calling "to get all the mujahideen together and defend this place." A few hours after the call went out, a Predator drone spotted two SUVs and a pickup truck headed towards the area. Fearing the vehicles contained insurgents heading to attack the American forces, the drone team—a remote pilot, sensor operator, and mission intelligence coordinator, working out of Creech Air Force Base in Nevada—began monitoring the convoy. Following the Predator team's assessment that the vehicles likely contained over 20 fighters and possibly a "high-value individual," commanders ordered nearby helicopters to attack. Three missiles hit the convoy, killing 23 civilians, including two boys under five years old, and wounding an additional nine adults and three children.[56]

An internal investigation conducted by United States Central Command (CENTCOM) faulted the Predator operators for providing intelligence that turned out to be inaccurate, as well as Special Operations Command for insufficiently supervising the operation. The report details discussions among the Predator's operating team, which show they interpreted ambiguous observations using assumptions that fit the preconceived narrative of insurgents heeding the call to fight the American forces near Khod. Using infrared cameras and grainy video, they spotted over 30 people, but misidentified some women and children as "military aged males." One thought he saw a rifle, but the group was uncertain. At dawn, the vehicles stopped and many of the passengers got out to wash and pray. This is typical behavior for millions of Muslims in the morning, and for many Taliban fighters before battle, and the operators assumed the latter without sufficiently considering the former. Despite these uncertain observations, the intelligence sent to the helicopter crews identified 21 military aged males, at least three weapons, and did not mention the possibility of children. A post-attack assessment team found no weaponry among the bodies.[57]

Though the Predator team made incorrect assumptions based on a preconceived narrative—a tragic example of human error—it is also worth noting the mistake could have been averted with better ISR equipment. If, upon spotting the convoy, the United States deployed a Smart SWARM, the operators would have seen the group contained many women and children, and that none of the passengers carried weapons. With a close-up view and facial recognition software, they might have also been able to determine there were no high value individuals in any of the vehicles. With this more accurate information, there would have been no need to fill in substantial blanks with guesses derived from the assumption the convoy likely contained fighters heeding the call to defend Khod. In response, United States commanders could have ordered ground forces to intercept the vehicles and keep them away from the battle, or simply ignored them. With a strategy based on massing information at a point of interest before acting, rather than acting upon limited information as interpreted by a few analysts, the United States could have avoided this disastrous error, as well as others like it, such as an airstrike that mistakenly killed 30 people at a hospital operated by Doctors Without Borders in Kunduz, Afghanistan in October 2015.[58]

THE SMART SWARM AND ROBOTIC AUTONOMY

At this point, we should probably talk about Skynet. In the *Terminator* movies, the United States military creates Skynet, a powerful artificial intelligence system, and gives it control of all computerized military hardware, including stealth aircraft and nuclear weapons. The system becomes self-aware and turns on its human masters, using America's arsenal to kill billions and take over the planet. This is one of many science fiction stories in which computers and robots end up threatening human survival—*2001: A Space Odyssey* and *The Matrix* are among the most famous—contributing to a widespread wariness of military robotics, especially autonomous systems. When thinking about how a computer would fight a war and what sort of technological developments would help achieve information superiority, it is important to consider the potential downsides of autonomous robots.

While pop-culture driven fears of current or forthcoming unmanned systems choosing to rebel against humanity are hyperbolic, military strategists and roboticists have raised reasonable concerns about the autonomy of killer machines. Since the earliest targeting computers on bombers, machines have assisted humans with life-or-death decisions, and automated systems capable of killing on their own have been in use for at least three decades. In the 1980s, the United States Navy began using the Phalanx C-RAM system to protect ships. The Phalanx, Patriot missile batteries, Israel's Iron Dome, and other anti-air and missile defense systems at sea or on land are capable of autonomously detecting enemy aircraft or incoming projectiles and firing without human input. Given the speed of missiles, the quickness of automated reactions has saved lives and equipment. However, on rare occasions, the systems have mistakenly identified targets and destroyed civilian or friendly aircraft, such as when the USS *Vincennes* mistakenly shot down Iran Air 655 on July 3, 1988, killing all 290 civilians on board.[59] This precedent creates concern that human beings remain "in the loop" and retain control of most decisions to fire weaponry as unmanned systems become increasingly capable of autonomous decision making.[60]

However, robots that can make decisions on their own provide numerous advantages, making increased autonomy inevitable. Besides the combat benefits associated with rapid decision making, autonomous machines act as a force multiplier. It is difficult for humans to operate more than one robot at a time, but autonomy enables multiple robots in the field per human operator. Additionally, if robots can carry out their primary functions without directions from a remote control, they are immune to jamming or unintentional signal interference. While many remain hesitant to grant robots the ability to select targets or choose when to fire without direct human input, if unmanned systems can demonstrate a near-perfect rate of success, most will come to accept it, much as virtually no one fears the use of autopilot in passenger jets or denounces automated anti-air defenses.

Even as individual robotic systems become more autonomous, they will not be taking over military strategy in the foreseeable future. It is dangerous to say technology will never be able to accomplish something—never is an awfully long time—but strategy is so complex, and must take into account so many variables, as to be far beyond the capabilities of early twenty-first-century computers. At the highest level, military strategy is linked to political objectives, which are definitionally dependent upon human preferences. However, even with human-determined objectives as inputs, the world's most powerful computers still could not handle military strategy.

Game-playing computers originally mastered tic-tac-toe and checkers, because there are a limited number of possible moves and situations. There are a lot more in checkers, but in both games there exists a perfect strategy that will either win or draw against every possible opponent.[61] Similarly, chess has a finite number of situations, albeit exponentially more than checkers. The upper bound of the number of possible arrangements of chess pieces has been mathematically proven to be, at most, $10^{46.25}$, and is probably lower by a few orders of magnitude.[62] When playing chess, computers play out millions of potential games from a given point and then select the move that leads to the highest probability of victory. Though no computer program has found an unbeatable strategy for chess like those for tic-tac-toe and checkers, given the finite number of possible positions, a sufficiently powerful computer could theoretically play every possible game and develop a formula for perfect chess.

Since IBM's Deep Blue first defeated grandmaster Gary Kasparov in a series of games in 1997, chess programs have had a strong record against humans, and in 2016 a program beat a human champion in the ancient Chinese game of go,[63] but computers have had more difficulty with poker. Programs have proven adept at simple versions of the game, with just two players and narrow betting limits, but have trouble determining optimal strategy in no-limit games where bettors may risk any or all of their chips at any point. Furthermore, each additional player increases the factors a computer must take into account. In January 2017, a program defeated four professional poker players in a series of one-on-one "heads up" no-limit contests,[64] but no machine has proven successful in the large, multi-player tournaments that determine world champions.[65]

Unlike chess, poker includes both randomness (the cards each player is dealt) and considerable unknowns. What cards do opponents have? If they bet, does it reflect the strength of their hand or are they bluffing? What type of strategy do they prefer, and have they changed strategies since the previous hand? How has their recent performance affected their mood? And how would they answer all of these questions about me? These factors create many more possible situations than chess, more than any current computer program can handle.

War includes many more pieces than chess, and far more unknowns than poker. Though artificial intelligence is advancing, as demonstrated by programs' success at go and heads up poker, the intelligence remains bounded, mastering a specific challenge with limited options. This indicates computers will increasingly excel at narrow military tasks, but they will not be taking over strategy any time soon.

With this in mind, the goal of an information-focused strategy of robotic warfare is to reduce unknowns to improve the decision making of the human participants. By massing informational capabilities to provide soldiers and commanders with the outline of outdoor and indoor areas and the location of people and relevant objects, a Smart SWARM could mitigate terrorists and insurgents' informational advantages. However, this system could not determine political objectives, guess opponents' strategies or enemy fighters' intentions, assess civilians' allegiances, or handle any of the other complex human considerations incorporated into strategy. Though the concept of the Smart SWARM draws upon ideas of how a computer would theoretically fight a war, it is a tool to assist human decision makers rather than a replacement for them.

Human–machine combinations are likely superior to either humans or machines alone. For example, as of 2017, the world's preeminent chess player is not a human grandmaster, nor a computer program, but a combination of the two the chess world calls a "centaur." Human players, utilizing multiple computer programs for recommendations but making the final decision on each move themselves, first defeated expert computers in 2005. In 2014, centaurs proved superior to the most sophisticated programs in tournaments that allowed both to participate.[66] At some point, with powerful enough computers, someone will likely develop the formula for perfect chess, making chess centaurs obsolete. However, there will never be a formula for perfect war.

As Deputy Secretary of Defense Robert Work argues, military centaurs can provide powerful advantages by combining the strengths of humans and machines.[67] The Smart SWARM is thus an example of what Work refers to as "human–machine teaming," since it leverages the capabilities of robots and computers to provide soldiers and commanders with enhanced awareness, facilitating faster and more accurate human decision making. ISR robots act autonomously to ensure coverage of a targeted area, while the central computer autonomously organizes the streams of information and identifies potential objects or persons of interest. Humans also input information into the network, and human decision makers utilize both human and robot-gathered information to help the Smart SWARM provide a more relevant understanding of the battlespace by determining where to mass informational capabilities. Human commanders then use this information to give smarter orders, while personnel in the field use the information to act more efficiently. The humans and machines mutually reinforce each other, combining robotic ISR capabilities with human judgment. Like other centaurs, it is greater than the sum of its parts. Together, people and robots can achieve information superiority.

Since the SWARM would only be gathering and processing information, rather than destroying property or killing people, it raises fewer concerns than weaponized robots about whether humans are sufficiently in the loop. It is likely the United States military will eventually utilize unmanned systems that decide on their own what and when to attack, but the Smart SWARM's function would be the same whether humans or robots are firing upon targets. Therefore, the system could provide an avenue for the development of military robotics that raises fewer concerns about autonomy. The SWARM enhances informational, rather than destructive, capabilities. It not only avoids ethical questions regarding whether machines or humans should be responsible for decisions to fire weapons, but also improves the ability of human soldiers and commanders to remain in the loop, by keeping them informed, in real-time, of the activities of both friendly and enemy humans and robots during combat.

INFORMATION-FOCUSED WARFARE

Information-gathering robots networked to powerful computers could provide information superiority and bring the ideal of Network-Centric Warfare closer to reality. NCW did not fully succeed in Iraq or Afghanistan because the information feeding into the network was of insufficient quality and quantity. In particular, asymmetric warfare against insurgencies appeared poorly suited for a strategy based on networking forces. Insurgent strategy relies on superior local knowledge, the ability to surprise, and exploitation of counterinsurgents' responsibility to protect everything at once. If utilized in a comprehensive informational strategy, robots have the potential to neutralize many of these advantages.

As described above, a Smart SWARM could facilitate a strategy based on massing ISR capabilities in which the United States would first know, and then act. By rapidly understanding a given battlespace, the United States could improve speed and efficiency while reducing error. It is important to note that the Smart SWARM strategy, like any updated strategy taking advantage of new technologies, does not require scrapping established military principles. For example, massing combat power would remain an essential element of warfare. However, by massing informational capabilities first whenever possible, the United States could achieve a greater economy of force, as increased awareness both increases the combat power of individual assets and decreases uncertainty regarding how much force is necessary to complete a designated task. Most of the technology necessary to create a Smart SWARM already exists, and, given current trends, it is reasonable to predict any additional elements will be available in the near future, especially if the Defense Department makes it a funding priority.

Though this chapter, like the rest of this book, focuses on counterinsurgency and other elements of asymmetric warfare, the central principles of the Smart SWARM strategy apply to the more symmetric challenges posed by state adversaries as well. With enough ISR assets around the world, complemented by monitors in cyberspace, a multilayered surveillance pyramid could provide global awareness; and rapidly massing regional or local informational capabilities as necessary would allow the United States to understand a situation before determining a course of action. However, achieving information superiority relative to near-peer or rogue state adversaries would require additional efforts to neutralize opponents' informational capabilities while protecting American ISR assets from physical or electronic attack.

CONCLUSION

Anticipating Future Risks

The twenty-first century will continue roboticizing, changing strategies of asymmetric warfare for both sides. Unmanned systems operating in the air, as well as on land or at sea, provide technologically advanced states a method of observing or attacking targets without risking human personnel, and gathering more information than humans could alone. Drones also allow terrorists and insurgents to observe and attack targets with greater anonymity. Governments cannot, and should not, fully control drone development, but they can anticipate risks, and work to make worst-case scenarios less likely.

Military strategies involving drones will proliferate as the technology spreads, with countries following the United States, and now United Kingdom's lead with extrajudicial drone strikes. This will include countries less friendly to America than Britain, such as Russia, China, and Iran. Therefore, it is in the United States' interest to establish an international legal architecture now, before the window of drones as a primarily American weapon finishes closing. America would have to voluntarily restrict its behavior somewhat, but establishing the laws and norms that restrict others' behavior and shape drone development would be a worthwhile tradeoff.

Robotic development is following a pattern similar to computers, and as the PC stage continues, and we enter the smart phone stage, commercially available UAVs will change insurgents' strategies, and states need to adapt accordingly. Already, for the first time in history, insurgents have aerial surveillance, aiding their ability to plan battles and coordinate forces. States need to factor this into their counter-insurgency strategies, and, to establish information dominance, must work to deny insurgents ISR capabilities.

As drones become more commonplace in cities, they will become less conspicuous, making it easier for a terrorist to case or attack a target. Electronically repelling drones from restricted airspace, or forcing them to land, would protect landmarks, airports, government buildings and other potential targets without excessively restricting individuals and businesses from using drones for film, photography, delivery, and whatever else people come up with in the future. Of course, someone will find a way around any anti-drone technology, and states should consider requiring drone manufactures to leave commercial models open to electronic defense signals. This would not stop everyone—someone would surely figure out how to modify their drone, and post the information online—but it would make it harder for terrorists to adapt commercially available UAVs to military purposes.

Autonomy enables faster decision making and more robots in the field per human operator, which makes further development a near certainty. While killer robots that make more of their own decisions might prompt discomfort at first, more people will accept them if they can demonstrate a track record of success. However, states should keep in mind asymmetric warfare is a political contest, and the political ramifications of an autonomous drone killing innocents, or somehow going haywire, would be considerable. Therefore, they should keep humans visibly in the loop, except for tasks where autonomy is necessary, most notably in the case of swarms.

Humans will issue orders to drone swarms, but each robot within the swarm must operate autonomously for the whole to be greater than the sum of its parts. Autonomous swarms could provide a comprehensive picture of a given battlespace that updates in real time, improving situational awareness and helping advanced militaries achieve information dominance. Though this book recommends focusing development on information-gathering swarms, combat swarms are likely in the future due to the unique capabilities they could provide. It is therefore essential to anticipate what types of group behavior could emerge from the rules governing each robot's actions.

AVOIDING A ROBOTIC FLASH CRASH

In the middle of the afternoon on May 6, 2010, America's main stock indices collapsed. In approximately 20 minutes, the Dow lost 9% of its value, wiping out more than a trillion dollars of wealth.[1]

Government regulators reversed thousands of trades, and the market quickly recovered, but this event, known as the "flash crash," demonstrates how actions that make sense for individual automated programs can lead to disastrous group behavior.

Automated trading programs did not trigger the crash, but they played a major role in making the market drop so far so quickly.[2] Government, academic, and private financial analysts have debated the extent to which the precipitating event was manipulation of the derivatives market, unusually large bets in a single direction, computer error, or something else. But whatever the original cause, the result was wild swings in the prices of some large companies' stocks, a spike in trading volume, and rapid drop in the major indices; and that happened because trading algorithms reacted to the erroneous information before anyone understood what was going on.

In 2010, trades conducted by computer algorithms accounted for over 80% of market volume.[3] Many of these programs sell when the price drops and buy as it rises, often in fractions of a second in response to tiny changes. For an individual trading algorithm, this rule is reasonable, based on the theory that falling stocks are likely to decline further, while rising securities will probably keep going up, at least in the short term. In the flash crash, the possibly erroneous price changes prompted more programs to sell than buy. Looking to unload stocks but unable to find buyers, they offered lower sale prices, causing further declines, triggering additional selling by additional programs. While trading based on stocks or indices' momentum might be profitable for an individual program, the collective decision that emerged from many programs acting on that rule was panic selling, leading to market collapse.

Unlike financial asset trades, some military actions cannot be undone. It is therefore crucial to create rules for individual robots in swarms that prevent disastrous collective decisions from emerging. For example, here's a seemingly logical rule for combat swarms: if a member of the swarm gets attacked, go help. That rule would facilitate rapid reactions that could quickly respond to surprise attacks and lead robotic forces to concentrate capabilities at weak points during combat. However, it could also cause a robotic flash crash.

For example, if members of two militaries' swarms accidently collide midair, or one misinterprets another's actions as threatening, multiple swarms could engage each other before anyone realizes what is going on. Besides risking an international incident, this could create a midair battle that puts anyone below in danger or quickly spirals into attacks against nearby military personnel. An accidental crash, misread of a threat, or even computer error could lead drones to mistakenly shoot down a passenger plane, or kill civilians on the ground.

To reduce the probability of this potential disaster, drone makers and militaries must carefully design rules governing autonomous drone swarms, especially those with combat capabilities, and run extensive simulations to ensure the right group behavior emerges from individual robots' decision making. Like financial regulators, states deploying drone swarms should be able to suspend activity if they notice something odd occurring. However, by the time human commanders realize what the swarm is doing, it may already have caused substantial damage.

With their ability to gather and process massive amounts of data, make split-second decisions, and facilitate a networked whole greater than the sum of its parts, drone swarms have immense potential. Maximizing that potential, and minimizing the risks, requires crafting rules governing individual robots' behavior that lead to ideal group decision making.

Notes

INTRODUCTION

1. Watson, "The Drones of ISIS."
2. "Drone Wars."
3. "A Wedding that Became a Funeral."
4. See Finkelstein, "Military Robotics: Malignant Machines or the Path to Peace?" pp. 5–6, for a more technical version of this definition.
5. Rawnsley, "It's a Drone's World. We Just Live in it."
6. "World of Drones: Military."
7. Ibid.
8. "List All Manufacturers."
9. Ying, "DJI sees jump in revenue."
10. "Drone maker DJI rumored to be planning for an IPO in 2017."
11. "Phantom."
12. Gallagher, "France cries foul at World Cup 'spy drone.'"
13. Moye, "ENIAC: The Army-Sponsored Revolution."
14. "ARPAnet."
15. "World Wide Web."
16. Weimann, "Terror on the Internet," p. 15.

CHAPTER 1   HOW TO FIGHT AN UNFAIR WAR

1. "Putin meets angry Beslan mothers."
2. "HM announces measures to enhance security."
3. "UCDP Conflict Encyclopedia."
4. Crawford, "War-related Death, Injury, and Displacement in Afghanistan and Pakistan 2001–2014."
5. "IAEA: Syria tried to build nuclear reactor."
6. Vick, "Spy Fail: Why Iran Is Losing Its Covert War with Israel."
7. "Attacks on Israeli civilians by Palestinians."

8. Giap, *People's War People's Army*, p. 103.

9. Al Muqrin, *Al-Qaida's Doctrine for Insurgency*, p. 92.

10. Guevara, *Guerrilla Warfare*, p. 20.

11. See Mao, "The Struggle in the Chingkang Mountains," 1928; "Problems of Strategy in China's Revolutionary War," 1936; "Problems of Strategy in Guerrilla War against Japan," 1938; "On Protracted War," 1938.

12. Mack, "Why Big Nations Lose Small Wars," p. 176.

13. Political scientists find this claim has strong empirical support. As Ivan Arreguin-Toft demonstrates, of 173 asymmetric wars from 1800–2000 (with more than 1,000 battle deaths in which the combatants have a measurable material power ratio of 5:1 or greater), the weaker side won only 23.2% of conflicts when directly confronting their stronger opponents, but defeated more powerful adversaries in 63.6% of confrontations in which they took an indirect approach and refused to fight on the stronger side's terms.

14. Thomas Schelling, in *Arms and Influence*, defined "compellence" as using a limited amount of force to convince an opponent (i.e. "compel" them) to abandon a particular behavior. "The threat that compels rather than deters," Schelling writes, "requires that the punishment be administered until the other acts, rather than if he acts," p. 70.

15. Mao, "On Protracted War," p. 208.

16. Mack, p. 184.

17. Ibid., p. 186.

18. Ibid., p. 187.

19. Mao, "On Protracted War."

20. Mack, "Why Big Nations Lose Small Wars," p. 177.

21. Mao, "Problems of Strategy in Guerrilla War Against Japan."

22. "Beirut Barracks Attack Remembered"; Khoury, "Last French peace-keepers ready to leave Beirut."

23. Kissinger, "The Vietnam Negotiations," p. 214.

24. From Keck and Sikkink's definition of a transnational advocacy network in *Activists Beyond Borders*, ch. 1.

25. Meigs, "Unorthodox Thoughts about Asymmetric Warfare," p. 8.

26. Arquilla and Ronfeldt, *Networks and Netwars*, p. 12.

27. Meigs, "Unorthodox Thoughts about Asymmetric Warfare," p. 10.

28. Eilstrup-Sangiovanni and Jones, "Assessing the Dangers of Illicit Networks."

29. Ibid., p. 21.

30. "Letter from al-Zawahiri to al-Zarqawi."

31. Galula, *Counterinsurgency Warfare*, p. 7.

32. See Galula, "Insurgency is cheap, counterinsurgency is costly," in *Counterinsurgency Warfare*, pp. 6–7.

33. "The 9/11 Commission Report."

34. Trinquier, *Modern Warfare: A French View of Counterinsurgency*, p. 23.

35. Valentino, Huth, and Balch-Lindsay, "Draining the Sea."

36. Kalyvas, "The Paradox of Terrorism in Civil War."

37. Crenshaw, *Terrorism in Context*, p. 4.
38. See Norton, *Hezbollah*, especially ch. 5, or Levitt, *Hamas: Politics, Charity, and Terrorism in the Service of Jihad*.
39. "The Warsaw Ghetto Uprising," The United States Holocaust Museum Online.
40. "Counterinsurgency Field Manual," the US Army and Marine Corps, foreword.
41. Dunlap, "We Still Need the Big Guns."
42. Valeriano and Bohannan, *Counter-Guerrilla Operations: The Philippine Experience*, p. 161.
43. Cronin, "How al-Qaida Ends," p. 42.
44. "No Military Solution to Iraq, U.S. General Says."
45. See Kydd and Walter, "The Strategies of Terrorism," for a discussion of the various signals sent by terrorist attacks.
46. Non-combatants are anyone other than actively engaged military. This definition of terrorism draws upon, but differs somewhat from, those offered by Bruce Hoffman (see *Inside Terrorism*, pp. 39–40) and Brigitte Nacos (see *Mass-Mediated Terrorism*, pp. 24–28).
47. Nacos, *Mass-Mediated Terrorism*, p. 14.
48. See Abrahms, "Why Terrorism Does Not Work."
49. Hanna, Martinez and Deaton, "ISIS publishes photo of what it says is bomb that downed Russian plane."
50. Ritter and Cook, "Death toll rises to 130 following Paris attacks."
51. "Islamic State and the crisis in Iraq and Syria in maps."

CHAPTER 2   THE ROBOTICS REVOLUTION

1. See "Counterinsurgency Field Manual."
2. Lyall and Wilson, "Rage Against the Machine."
3. Nagl, *Learning to Eat Soup with a Knife*.
4. Jeffrey, "Why Counterinsurgency Doesn't Work."
5. SIPRI Military Expenditure Database.
6. "Defense Budget Priorities and Choices."
7. "Fiscal Year 2016 Budget Request," ch. 2, pp. 2–3.
8. "Defense Budget Priorities and Choices," p. 10.
9. "Fiscal Year 2016 Budget Request," pp. 8–16.
10. Pincus, "Air Force to Train More Remote than Actual Pilots."
11. Mueller, *War, Presidents, and Public Opinion*.
12. Data an average of polls from Gallup, CNN and Opinion Research Corporation, and Fox News and Opinion Dynamic. See: Wayner, "American Approval Rating (Percent) of War in Afghanistan."
13. Data from "Operation Enduring Freedom" at *iCasualties.org*.
14. "Afghanistan," *Gallup*.
15. "Washington Post-ABC News Poll," February 4, 2012.

16. "Drone Strikes Widely Opposed, Global Opinion of Obama Slips, International Policies Faulted," *Pew Global Attitudes Project*.
17. "AP–GfK Poll: Americans approve of drone strikes on terrorists."
18. "Public Continues to Back U.S. Drone Attacks."
19. Singer, "Do Drones Undermine Democracy?"
20. Atwood and Klein, "Vecna's Battlefield Extraction-Assist Robot BEAR"; "High Performance Hydraulics for Industrial Applications."
21. Ruppert, "Battlefield Extraction-Assist Robot to Rescue Wounded on Battlefield."
22. "BigDog Overview."
23. "BigDog – The Most Advanced Rough-Terrain Robot on Earth."
24. "Dynamic Robot Manipulation."
25. Diaz, "Robot horse gets first taste of real-world action with the US Marines."
26. Pfeiffer, "DARPA Unveils Robotic Mule."
27. Smith, "Marines send its 'AlphaDog' robot to the farm."
28. "Ground Robots – 510 PackBot."
29. "Operation Enduring Freedom," iCasualties.org.
30. "iRobot PackBot 510 with Engineer Kit."
31. "PackBot Tactical Robot."
32. "Operation Enduring Freedom," *iCasualties.org*.
33. "Talon Specifications."
34. "Armed, Aware and Dangerous."
35. Singer, *Wired for War*, p. 38.
36. "Counter Rocket, Artillery, and Mortar (C-RAM)."
37. "MK 15 – Phalanx Close-In Weapons System (CIWS)."
38. "Counter Rocket, Artillery, and Mortar (C-RAM)."
39. Singer, *Wired for War*, p. 38.
40. "Counter Rocket, Artillery, and Mortar (C-RAM)."
41. "A Laser Phalanx?"
42. "Raytheon's Mobile Land-Based Phalanx Weapon System Completes Live-Fire Demonstration."
43. "NBS MANTIS Air Defense Protection System, Germany."
44. Szondy, "Neither rain, nor fog, nor wind stops Boeing's laser weapon destroying targets."
45. "A Laser Phalanx?"
46. Thompson, Mark, "Iron Dome: A Missile Shield That Works."
47. Singer, *Wired for War*, p. 30.
48. "Armed Robots March into Battle."
49. Singer, *Wired for War*, p. 31.
50. "PETMAN – BigDog Gets a Big Brother."
51. Ibid.
52. Ackerman, "ATLAS DRC Robot is 75 Percent New, Completely Unplugged."
53. Yirka, "Makers of infamous BigDog robot unveil human version."
54. "PETMAN – BigDog Gets a Big Brother."

55. Guizzo and Ackerman, "How South Korea's DRC-HUBO Robot Won the DARPA Robotics Challenge."
56. Sofge, Erik, "The DARPA Robotics Challenge Was a Bust."
57. Rawnsley, "Darpa's Cheetah-Bot Designed to Chase Human Prey."
58. "Cheetah Robot runs 28.3 mph, a bit faster than Usain Bolt."
59. "Unmanned Aerial Vehicles."
60. Wagner, *Lightning Bugs and Other Reconnaissance Drones*.
61. Singer, *Wired for War*, p. 56.
62. "Predator RQ-1/MQ-1/MQ-9 Reaper – United States of America."
63. "MQ-1 Predator."
64. Coll, *Ghost Wars*, p. 529.
65. Pitzke, "How Drone Pilots Wage War."
66. "Predator UAS."
67. "Predator RQ-1/MQ-1/MQ-9 Reaper – United States of America."
68. Grey, "U.S. Kills al Qaeda Leaders by Remote Control."
69. "CIA 'killed al-Qaeda suspects' in Yemen."
70. Quoted in "'Reaper' moniker given to MQ-9 unmanned aerial vehicle."
71. "Predator B UAS."
72. "Predator C Avenger."
73. Carey, "General Atomics Predator C Avenger ER Makes First Flight."
74. Whitlock, "More Air Force drones are crashing than ever, as mysterious new problems emerge."
75. Dobbing and Cole, "Israel and the Drone Wars," p. 10.
76. "MQ-1B Predator"; Alex, "IAI Heron/Machatz-1 Unmanned Aerial Vehicle."
77. Dobbing and Cole, "Israel and the Drone Wars," p. 10.
78. Ibid., pp. 12–13.
79. Perry and Williams, "Israeli drone strike in Syria kills two near frontier: Hezbollah's al-Manar TV."
80. "UK Drone Strike Stats."
81. Yeung, "UK air strikes kill 1,000 Isis fighters in Iraq and Syria but no civilians, officials claim."
82. Patel, "War on Isis: 305 jihadists killed by RAF drones in Iraq with 'no known civilian casualties.'"
83. Watt, "The 'kill list': RAF drones have been hunting UK jihadis for months."
84. "F/A-18 *Hornet* Strike Fighter."
85. "F-22 Raptor."
86. "Producing, Operating and Supporting a 5th Generation Fighter."
87. Axe, "Buyer's Remorse: How Much Has the F-22 Really Cost?"
88. "MQ-9 Reaper."
89. Pardesi, "Unmanned Aerial Vehicles/Unmanned Combat Aerial Vehicles: Likely Missions and Challenges for the Policy- Relevant Future."
90. "Dogfight between MQ-1 Predator drone and MiG-25 Foxbat."
91. Axe, "Predator Drones Once Shot Back at Jets … But Sucked At It."
92. "Unmanned Aircraft Systems Flight Plan."

93. Terdiman, Daniel, "Drone dogfights by 2015? U.S. Navy preps for futuristic combat."
94. "Switchblade – Miniature Loitering Weapon."
95. Hennigan, "Pentagon to soon deploy pint-sized but lethal Switchblade drones."
96. Dunnigan, "Switchblade Enters Service."
97. "Switchblade – Miniature Loitering Weapon."
98. "UAS Advanced Development: Switchblade."
99. Dunnigan, "Switchblade Enters Service."
100. "US Military Bringing a Switchblade to A Gun Fight."
101. Hennigan, "Pentagon to soon deploy pint-sized but lethal Switchblade drones."
102. Dunnigan, "Switchblade Enters Service."
103. Ibid.
104. "U.S. Army Awards AeroVironment $5.1 Million Order for Switchblade Loitering Munition Systems and Services."
105. "U.S. Army increases orders for SwitchBlade microUAV based guided weapons."
106. "Robocopter arrives."
107. Davies, "The Marines' Self-Flying Chopper Survives a Three-Year Tour."
108. "Unmanned K-MAX Wins Top Honors, USMC Praise."
109. McLeary, "Marines extend Afghan deployment of cargo UAV."
110. McLeary, "K-MAX Chugging Along in Afghanistan."
111. Davies, "The Marines' Self-Flying Chopper Survives a Three-Year Tour."
112. Sanborn, "Beacon improves UAVs cargo-delivery accuracy."
113. "Robocopter arrives."
114. Shachtman, "Flying Spy Surge: Surveillance Missions Over Afghanistan Quadruple."
115. Wagner, *Lightning Bugs and Other Reconnaissance Drones*, pp. xi, xii.
116. Ibid., p. 208.
117. "RQ-4 Global Hawk."
118. "What Is Synthetic-Aperture Radar?"
119. Lum, "The Measure of MASINT."
120. "RQ-4 Global Hawk," p. 4.
121. Singer, *Wired for War*, p. 36.
122. "RQ-4 Global Hawk", p. 4.
123. Ibid., p. 2.
124. "Lockheed Martin RQ-170 Sentinel Unmanned Aerial Vehicle."
125. Dsouza, "RQ-170 Sentinel 'Beast of Kandahar.'"
126. "Lockheed Martin RQ-170 Sentinel Unmanned Aerial Vehicle."
127. "Gorgon Stare."
128. Nakashima and Whitlock, "With Air Force's Gorgon Drone 'we can see everything.'"
129. "Gorgon Stare."
130. Nakashima and Whitlock, "With Air Force's Gorgon Drone 'we can see everything.'"

131. Clark, "Gorgon Stare Blinks A Lot."
132. Ibid.
133. Beizer, "BAE to Develop Surveillance System."
134. Gallagher, "Could the Pentagon's 1.8 Gigapixel Drone Camera Be Used for Domestic Surveillance?"
135. "Autonomous Real-Time Ground Ubiquitous Surveillance-Imaging System (ARGUS-IS)."
136. Gallagher, "Could the Pentagon's 1.8 Gigapixel Drone Camera Be Used for Domestic Surveillance?"
137. "Integrated Sensor Is Structure (ISIS)."
138. Hoffman, "PBS Features DARPA'S ARGUS-IS."
139. Trimble, "Lockheed Martin to Build the Mother of All Airborne Radars."
140. Warwick, "Airship Programs – Not So Buoyant, Says GAO."
141. Hoffman, "PBS Features DARPA'S ARGUS-IS."
142. Nakashima and Whitlock, "With Air Force's Gorgon Drone 'we can see everything.'"
143. Shachtman, "Flying Spy Surge: Surveillance Missions Over Afghanistan Quadruple."
144. Vickery, "Operation Inherent Resolve: An Interim Assessment."
145. Nakashima and Whitlock, "With Air Force's Gorgon Drone 'we can see everything.'"
146. "UAS Advanced Development: Raven RQ-11A" and "UAS: Raven RQ-11B."
147. "RQ-11 Raven Unmanned Aerial Vehicle."
148. Singer, "Wired for War," p. 37.
149. "RQ-11 Raven Unmanned Aerial Vehicle."
150. "Why Soldiers Hate the Raven UAV."
151. Singer, "Wired for War," p. 37.
152. "Wasp III."
153. "UAS: Wasp AE."
154. "Wasp III."
155. Hill, "Toy-Size Helicopter Drones Now on Surveillance Duty in Afghanistan."
156. "PD-100 PRS – Your Personal Reconnaissance System."
157. Hill, "Toy-Size Helicopter Drones Now on Surveillance Duty in Afghanistan."
158. "PD-100 PRS – Your Personal Reconnaissance System."
159. Quoted in "Miniature surveillance helicopters help protect front line troops."
160. "PD-100 PRS – Your Personal Reconnaissance System."
161. Hoffman, "British soldiers flying nano helicopters in Afghanistan."
162. "Miniature surveillance helicopters help protect front line troops."
163. Greenberg, "Flying Drone Can Crack Wi-Fi Networks, Snoop on Cell Phones."
164. "About Us," *the Rabbit-Hole*.

165. Humphries, "WASP: The Linux-powered flying spy drone that cracks Wi-Fi & GSM networks."
166. "The Throwbot XT with Audio Capabilities."
167. "The Military's New Weapon: Mini Spy Robots You Throw Like Grenades."
168. "Army Orders 1,100 Recon Scout XT Robots from ReconRobotics."
169. "The Throwbot XT with Audio Capabilities."
170. "Decibel Levels of Everyday Sounds."
171. Crane, "Anti-Sniper/Sniper Detection/Gunfire Detection Systems at a Glance."
172. Sofge, "5 Robots We Should Deploy Right Now," p. 4.
173. Ibid.
174. Ibid.

CHAPTER 3   DRONE STRIKES

1. "Could The Use Of Flying Death Robots Be Hurting America's Reputation Worldwide?"
2. Brooks, "Drones and the International Rule of Law."
3. "Drone Wars."
4. Allen, "WikiLeaks: Yemen Covered Up US Drone Strikes."
5. "Covert Drone War."
6. "Year of the Drone."
7. Low percentage of civilian deaths calculated using the low estimated total deaths and low estimated militant death figures. High percentage of civilian deaths calculated using the high estimated total and high militant death figures. As a result, the percentage of civilian deaths calculated by using the low estimates is often greater than that calculated by using the high estimates.
8. Low percentage of civilian deaths calculated using the low estimated total deaths and low estimated militant death figures. High percentage of civilian deaths calculated using the high estimated total and high militant death figures. As a result, in some years, the percentage of civilian deaths calculated by using the low estimates is greater than that calculated by using the high estimates.
9. Khan, "Pakistani Taliban: US Drone Strikes Forcing Militants Underground."
10. Bergen and Rowland, "Obama Ramps Up Covert War in Yemen."
11. "Obama's Covert War in Yemen."
12. "Covert Drone Wars."
13. DeYoung and Miller, "White House releases its count of civilian deaths in counterterrorism operations under Obama."
14. "Covert Drone Wars.".
15. "Final Vote Results for Roll Call 342."
16. "Text of Authorization for Use of Military Force."

17. Risen and Johnston, "Threats and Responses: Hunt for Al Qaeda; Bush Has Widen Authority of C.I.A. to Kill Terrorists."

18. Miller, "Plan for hunting terrorists signals U.S. intends to keep adding names to kill lists."

19. Becker and Shane, "Secret 'Kill List' Proves a Test of Obama's Principles and Will."

20. "Britain's jihadi kill list."

21. Watt, "The 'kill list': RAF drones have been hunting UK jihadis for months."

22. Botelho and Starr, "U.S. 'reasonably certain' drone strike killed ISIS mouthpiece 'Jihadi John.'"

23. Watt, "The 'kill list': RAF drones have been hunting UK jihadis for months."

24. "Britain's jihadi kill list."

25. Watt, "The 'kill list': RAF drones have been hunting UK jihadis for months."

26. Madhani, "Cleric al-Awlaki Dubbed 'bin Laden of the Internet.'"

27. Savage, "Secret U.S. Memo Made Legal Case to Kill a Citizen."

28. Cohen, "When can a government kill its own people?"

29. Cole, "Al-Awlaqi Should have been Tried in Absentia."

30. Shane and Schmitt, "One Drone Victim's Trail from Raleigh to Pakistan."

31. Van Dyk, "Who were the 4 U.S. citizens killed in drone strikes?"

32. Diamond, "U.S. drone strike accidentally killed 2 hostages."

33. Khatchadourian, Raffi, "Azzam the American."

34. Specia, "Who are the Americans who have been killed by U.S. drone strikes?,"

35. "AP–GfK Poll: Americans approve of drone strikes on terrorists."

36. Mullen, "Al Qaeda's second in command killed in Yemen strike; successor named."

37. Byman, "Can Al Qaeda in the Arabian Peninsula Survive the Death of Its Leader?"

38. "Yemen-based al Qaeda group claims responsibility for parcel bomb plot."

39. Mazzetti, Worth and Lipton, "Bomb Plot Shows Key Role Played By Intelligence."

40. Reidel, "Al-Qaida's Hadramawt emirate."

41. Greenfield, "The Case Against Drone Strikes on People Who Only 'Act' Like Terrorists."

42. Shane, "Drone Strikes Reveal Uncomfortable Truth: U.S. Is Often Unsure About Who Will Die."

43. Greenfield, "The Case Against Drone Strikes on People Who Only 'Act' Like Terrorists."

44. Guerin, "US drone war in Pakistan prompts fear and anger."

45. Greenfield, "The Case Against Drone Strikes on People Who Only 'Act' Like Terrorists."
46. Dozier, "Report: U.S. drone strike may have killed up to a dozen civilians in Yemen."
47. Though US government claims regarding civilian casualties from extrajudicial drone strikes deserve skepticism, since the United States has a clear incentive to downplay accidental damage to civilians, there is no clear incentive to portray one type of aircraft as more likely to harm civilians than another, making this data more reliable.
48. See: Dobrydney, David, "Combined Forces Air Component Command Airpower Statistics."
49. Ibid.
50. "Afghanistan: Annual Report 2012 Protection of Civilians in Armed Combat."
51. Bergen, "Secrets of the bin Laden treasure-trove."
52. "Video shows capture of Abu Anas al-Libi."
53. Weiser and Schmidt, "Qaeda Suspect Facing Trial in New York Over Africa Embassy Bombings Dies."
54. "U.S. Navy SEALs fail to capture al-Shabaab commander."
55. Bowden, Mark, "A defining battle."
56. Almosawa and Nordland, "U.S. Fears Chaos as Government of Yemen Falls."
57. "Fatalities in Terrorist Violence in Pakistan 2003–2015."
58. Hanna and Almasmari, "Al Qaeda freed 6 inmates in Yemen prison attack, officials say."
59. Booth and Black, "WikiLeaks cables: Yemen offered US 'open door' to attack al-Qaida on its soil."
60. "Yemen crisis: US troops withdraw from air base."
61. "Secret US drone base in Saudi Arabia revealed."
62. Miller and Woodward, "Secret memos reveal explicit nature of U.S., Pakistan agreement on drones."
63. Friedersdorf, "Yes, Pakistanis Really Do Hate America's Killer Drones."
64. "Foreign Assistance: Summary Tables Fiscal Year 2015."
65. Ferran, "Top US Spy: Intel Cooperation with Pakistan 'On the Upswing.'"
66. Gannon and Abbot, "Criticism alters US drone program in Pakistan."
67. "Saudi Arabia Says It Foiled Attack on U.S. Embassy, Arrested ISIS Supporters."
68. Whitlock, "Italy convicts 22 CIA operatives, U.S. Air Force colonel in rendition case."
69. Fishel, et al., "EXCLUSIVE: Undercover DHS Tests Find Security Failures at US Airports."
70. Bergen, "Time to declare victory: al Qaeda is defeated."
71. Andrews and Lindeman, "The Black Budget."

72. Gellman and Poitras, "U.S., British intelligence mining data from nine U.S. Internet companies in broad secret program."
73. Nakashima, "Officials: surveillance programs foiled more than 50 terrorist plots."
74. Bergen, "Time to declare victory: al Qaeda is defeated."
75. Friedersdorf, "Yes, Pakistanis Really Do Hate America's Killer Drones"; Murray, "Anger at US drone war continues in Yemen."
76. Mothana, "How Drones Help Al Qaeda"; Pilkington and MacAskill, "Obama's drone war a 'recruitment tool' for Isis, say US air force whistleblowers."
77. Quoted in Bergen and Rowland, "Obama Ramps Up Covert War in Yemen."
78. Mezzofiore, "Al-Qaeda in Yemen ideological leader Ibrahim al-Rubaish killed in drone strike."
79. "Yemen's al-Qaeda: Expanding the Base."
80. Reidel, "Al-Qaida's Hadramawt emirate."
81. Crowcroft, "Al-Qaeda seizes major airport and Mukalla oil terminal in Southern Yemen."
82. Crowcroft, "Yemen: Who is the new leader of al-Qaeda in the Arabian Peninsula Qasim al-Raymi?"
83. Weiss, "AQAP shows fighting in strategic Yemeni city in new video."
84. "Yemen's al-Qaeda: Expanding the Base."
85. Byman, "Do Targeted Killings Work?"; David, "Fatal Choices: Israel's Policy of Targeted Killing."
86. Hafez and Hatfield, "Do Targeted Assassinations Work? A Multivariate Analysis of Israeli Counter-Terrorism Effectiveness during Al-Aqsa Uprising."
87. Byman, "Do Targeted Killings Work?"
88. Norton, *Hezbollah*.
89. "Hezbollah's Rocket Force"; Stuster, "Why Hezbollah's New Missiles Are a Problem for Israel"; Issacharoff, "Israel raises Hezbollah rocket estimate to 150,000."
90. Cordesman, "The Lessons of the Israeli–Lebanon War."
91. "Lebanon rivals agree to crisis deal."
92. Jordan, "When Heads Roll: Assessing the Effectiveness of Leadership Decapitation."
93. Price, "Targeting Top Terrorists: How Leadership Decapitation Contributes to Counterterrorism."
94. Cronin, "How al-Qaida Ends: The Decline and Demise of Terrorist Groups."
95. Dorell, "Iranian support for Yemen's Houthis goes back years."
96. Reidel, "Al-Qaida's Hadramawt emirate."
97. Johnston, "Does Decapitation Work? Assessing the Effectiveness of Leadership Targeting in Counterinsurgency Campaigns."
98. Jordan, "When Heads Roll: Assessing the Effectiveness of Leadership Decapitation," p. 755.

99. Johnston and Sarbahi, "The Impact of US Drone Strikes on Terrorism in Pakistan."

100. Jordan, "The Effectiveness of the Drone Campaign against Al Qaeda: A Case Study."

101. Rassler et al., "Letters from Abbottabad: Bin Laden Sidelined?"

102. "The Al-Qaida Papers − Drones."

103. Jordan, "The Effectiveness of the Drone Campaign against Al Qaeda: A Case Study," pp. 13−14.

104. "Dead, captured and wanted."

105. Benson, "Is the core of al Qaeda on its last legs?"

106. Bergen, *The Longest War.*

107. Vardi, Nathan, "Is al Qaeda Bankrupt?"

108. Bialik, "Shadowy Figure: Al Qaeda's Size is Hard to Measure."

CHAPTER 4   TERRORISTS AND INSURGENTS, ARMED WITH DRONES

1. Shear and Schmidt, "White House Drone Crash Described as a U.S. Worker's Drunken Lark."

2. "Phantom," *DJI Innovations.*

3. "DJI Phantom," *Google Shopping.*

4. Shear and Schmidt, "White House Drone Crash Described as a U.S. Worker's Drunken Lark."'

5. "TM-43-0001-29 Army Ammunition Data Sheets for Grenades."

6. "Predator B UAS"; "Predator RQ-1/MQ-1/MQ-9 Reaper − United States of America."

7. Axe, "Predator Drones Once Shot Back at Jets... But Sucked At It."

8. "Switchblade − Miniature Loitering Weapon"; "UAS Advanced Development: Switchblade."

9. Hennigan, W.J., "United Arab Emirates set to buy U.S. Predator drones."

10. Binnie, "General Atomics confirms UAE Predator delivery."

11. Vardi, "Is al Qaeda Bankrupt?"

12. "What is the FATF?"

13. Figures from FY-2013 United States Department of Defense Procurement Budget. Cost includes the aircraft, ground controls station, modifications and payloads. See: "MQ-1 Predator/MQ-9 Reaper."

14. Vardi, "Is al Qaeda Bankrupt?"

15. Pagliery, "Inside the $2 billion ISIS war machine."

16. Thompson, Mark, "U.S. Bombing of ISIS Oil Facilities Showing Progress."

17. "Islamic State to halve fighters' salaries as cost of waging terror starts to bite."

18. "UAS: Wasp AE."

19. Barnett, "Hezbollah takes responsibility for last week's drone over Israel."

20. Harel, "Air Force: Hezbollah drone flew over Israel for five minutes."

21. "Mohajer (UAV)."

22. "Syrian Intelligence May Have Worked with Hizballah on UAV Launchings."
23. Ephron, "Hizbullah's Worrisome Weapon."
24. "Ababil (Swallow) Unmanned Aerial Vehicle."
25. Bergman, "Hezbollah boosting drone unit."
26. Harel et al., "Hezbollah drone brought down over Galilee held 30 kg of explosives."
27. Bergman, "Hezbollah boosting drone unit."
28. Cenciotti, "Hamas Flying An Iranian-Made Armed Drone Over Gaza."
29. "Hezbollah Drones Target Al-Nusra Front's Positions at Syrian Border."
30. Blanford, "Hizbullah airstrip revealed."
31. Cenciotti, "Iran's New Spy Drone Is an Israeli Hermes 450/Watchkeeper Clone Capable of Carrying Missiles."
32. Masi, "Hezbollah Allegedly Using Drones Against Al Qaeda In Battle For Qalamoun"; "Hezbollah strike Qalamoun jihadist with help of drone."
33. "Middle East Crisis: Facts and Figures."
34. Cordesman, "The Lessons of the Israeli–Lebanon War," p. 16.
35. "Nasrallah hits out at government."
36. Cordesman, "The Lessons of the Israeli–Lebanon War."
37. Ibid., p. 3.
38. "Middle East Crisis: Facts and Figures."
39. "Hezbollah's rocket force."
40. Schneider, "Hezbollah rearms away from border."
41. "Hezbollah Displays Iranian Fajr-5 Missile."
42. "SkyGrabber."
43. Gorman et al., "Insurgents Hack U.S. Drones."
44. "Researchers use spoofing to 'hack' a drone."
45. Mackenzie and Duell, "We hacked U.S. drone."
46. "Drone shot down over Iran 'lost' over Afghanistan last week."
47. "Researchers use spoofing to 'hack' a drone."
48. Kamkar, "SkyJack."
49. Szabo, "Let's hack a drone!"
50. Rodday, "Hacking a Professional Drone."
51. Greenberg, "Hacker says he can hijack a $35k police drone a mile away."
52. "Press Release – New FAA Rules for Small Unmanned Aircraft Systems Go Into Effect."
53. "Drone Laws By Country"; "French Drone Regulations – Updated as of 6/23/2015"; "Civilian drones and the legal issues surrounding their use."
54. Brandom, "How Brazil is trying (and failing) to keep drones away from the Olympics."
55. Roudik, "Russia: Commercial and Private Drones to Be Outlawed in Moscow."
56. "Drone Laws By Country."

57. "Teal Group Predicts Worldwide UAV Market Will Total $91 Billion in Its 2014 UAV Market Profile and Forecast."
58. "The Drones Report: Market forecasts, regulatory barriers, top vendors, and leading commercial applications."
59. "Drones: Reporting for Work."
60. Sasso, "Hollywood wants drones for filmmaking."
61. Wingfield and Sengupta, "Drones Set Sites on U.S. Skies."
62. Sasso, "Hollywood wants drones for filmmaking."
63. Teinowitz, "Hollywood to the FAA: Let Us Use Drones!"
64. Ibid.
65. "The Totally New SARAH Unmanned Aerial System."
66. "Move Like You Think, A Thought Made Invisible."
67. "Flying-Cam and Bond 007 'Skyfall.'"
68. "Official FAA Approval for Flying-cam 3.0 SARAH."
69. Janik and Armentrout, "Industry looks to use drones for commercial purposes."
70. Koebler, "Burrito Bomber Attacks Hunger with Drone-Delivered Mexican Food."
71. "Burrito Bomber."
72. "Amazon Unveils Futuristic Plan: Delivery by Drone."
73. Chung, "Amazon tests delivery drones at a secret site in Canada."
74. Lovelace, "The future arrives? Amazon's Prime Air completes its drone delivery."
75. "Amazon Prime Air."
76. Mac, "Amazon Unveils New Drone Design Ahead of Cyber Monday."
77. Mac, "Amazon Patent Reveals How Delivery Drones Could Avoid Crashing Into Your Home."
78. Elliott, "Do You Know Where Your Slogan Is?"
79. Bidgood, "Massachusetts Man Gets 17 Years in Terrorist Plot."
80. Johnson, "Man accused of plotting drone attacks on Pentagon, Capitol."
81. "E-Flite F-86 Sabre 15 Ducted Fan Jet ARF."
82. Cacace, "Affidavit of Special Agent Gary S. Cacace: 11-mj-4270-tsh."
83. Johnson, "Man accused of plotting drone attacks on Pentagon, Capitol."
84. Cacace, "Affidavit of Special Agent Gary S. Cacace: 11-mj-4270-tsh," pp. 39–40.
85. "Man, 26, charged in model airplane plot to bomb the Pentagon," p. 2.
86. Ibid., p. 1.
87. Anderson, "A newbie's guide to UAVs."
88. Ibid.
89. "ArduPilot."
90. "Download the Arduino Software."
91. Anderson, "A newbie's guide to UAVs."
92. Anderson, "The DIY Drones Mission (aka The Five Rules)," site policies.

93. Ibid., rule #3.
94. "ArduCopter User Group."
95. "Camera Restrictions in New York."
96. Geoghegan, Tom, "Innocent photographer or terrorist?"
97. "Solo."
98. Michelzon,"New Video Shows ISIS Using Drones to Plan Battles."
99. Cenciotti, "ISIS Surveillance Drone Is Only an Amateurish Remote-Controlled Quad-Copter."
100. Weiss, "Islamic State uses drones to coordinate fighting in Baiji."
101. Hall, "ISIS propaganda, Call of Duty Style."
102. Schmidt and Schmitt, "Pentagon Confronts a New Threat from ISIS: Exploding Drones."
103. Warrick, "Use of weaponized drones by ISIS spurs terrorism fears."
104. "U.S. aircraft strike ISIS drone in Iraq: officials."
105. "ISIS drone shot down by Peshmerga."
106. Fadel, "ISIS Drone Downed by the Syrian Army at Kuweires Airbase in Aleppo."
107. Wingfield, "New F.A.A. Report Tallies Drone Sightings, Highlighting Safety Issues"; "UAS Sightings Reports."
108. "Reports and Analysis."
109. "Causal Factors & Risk Ratings."
110. Lamden and Crossley, "Four planes have been involved in fatal near misses with drones at major British airports including Heathrow in the last month."
111. "'Drone' hits British Airways plane approaching Heathrow Airport."
112. Quinn, "UK police see spike in drone incidents."
113. Young, "Drones increasingly buzzing too close to Canadian airports: reports."
114. "Paris drones: New wave of alerts."
115. Andrews, "A commercial airplane collided with a drone in Canada, a first in North America."
116. Donohue and Goldiner, "How often do birds cause plane crashes?"
117. Starr and Shoichet, "Russian plane crash: U.S. intel suggests ISIS bomb brought down jet."
118. "Press Release – FAA Announces Small UAS Registration Rule."
119. "Registration and Marking Requirements for Small Unmanned Aircraft: A Rule by the Federal Aviation Administration."
120. Oswald, "This Anti-UAV Octocopter Uses a Ballistic Net Cannon to Disable Smaller Drones."
121. Matyszczyk, "Need to take down a drone? A munitions company offers firepower."
122. Golson, "Welcome to the World, Drone-Killing Laser Cannon"; Atherton, "Israeli Contractor Rafael Shows Off Anti-Drone Laser in Korea."
123. Ackerman, "Dutch Police Training Eagles to Take Down Drones."

CHAPTER 5   THE SMART SWARM STRATEGY

1. Cebrowski and Garstka, "Network-Centric Warfare: Its Origin and Future."
2. "The Implementation of Network-Centric Warfare."
3. "Joint Doctrine for Information Operations" (Joint Pub 3-13).
4. Alberts, Garstka, and Stein, *Network Centric Warfare: Developing and Leveraging Information Superiority*, p. 8.
5. Singer, *Wired for War*, pp. 193–194.
6. Ibid., p. 193.
7. Thompson, Loren, "The Twilight of Network-Centric Warfare."
8. Singer, *Wired for War*, p. 190.
9. Vego, Milan, "The NCW Illusion."
10. "Operation Iraqi Freedom," *iCasualties.org*.
11. Singer, *Wired for War*, pp. 208–12.
12. Ibid., p. 209.
13. Ibid., p. 210.
14. Ibid.
15. "Fido XT Explosives Trace Detector."
16. "PackBot Tactical Robot," *Defense Update*.
17. Hill, "Toy-Size Helicopter Drones Now on Surveillance Duty in Afghanistan."
18. Danigelis, "Tiny Dragonfly UAV Flies and Hovers to Spy."
19. Green, "Dragonfly Robotic Insect UAV is Freaking Cool."
20. Danigelis, "Tiny Dragonfly UAV Flies and Hovers to Spy."
21. Green, "Dragonfly Robotic Insect UAV is Freaking Cool."
22. "Robodiptera."
23. "Autonomous Flying Microrobots (RoboBees)."
24. "ArduPilot."
25. "RQ-4 Global Hawk: High-Altitude, Long-Endurance Unmanned Aerial Reconnaissance System."
26. "Center for Neural and Emergent Systems."
27. Simonite, "A Brain-Inspired Chip Takes to the Sky."
28. Sahin and Franks, "Measurement of Space: From Ants to Robots."
29. "Swarm-bots: Swarms of self-assembling artifacts."
30. "Swarmanoid: Towards Humanoid Robotic Swarms."
31. Trianni and Dorigo, "Emergent Collective Decisions in a Swarm of Robots."
32. Labella, Dorigo, and Deneubourg, "Division of Labor in a Group of Robots Inspired by Ants' Foraging Behavior."
33. Moubarak and Ben-Tzvi, "Modular and reconfigurable mobile robotics"; Patil, Abukhalil and Sobh, "Hardware Architecture Review of Swarm Robotics System: Self-Reconfigurability, Self-Reassembly, Self-Replication."
34. Stirling, et al. "Indoor Navigation with a Swarm of Flying Robots."
35. "Mini army drones developed."

36. Smalley, "LOCUST: Autonomous, swarming UAVs fly into the future."
37. Condliffe, "A 100-Drone Swarm, Dropped from Jets, Plans Its Own Moves."
38. Kharpal, "Amazon wins patent for a flying warehouse that will deploy drones to deliver parcels in minutes."
39. See, for example, "AutoCAD Map 3D."
40. Dokmanic, et al. "Acoustic echoes reveal room shape."
41. "Tiny airplanes and subs from University of Florida laboratory could be next hurricane hunters."
42. Bittel, "Studying Hurricanes With Swarms of Smart Drones."
43. Sonka et al., *Image Processing, Analysis, and Machine Vision.*
44. Fingas, "Google lands patent for automatic object recognition in videos, leaves no stone untagged."
45. Markoff, "How Many Computers to Identify a Cat? 16,000."
46. Markoff, "Researchers Announce Advance in Image-Recognition Software."
47. Davey, "Explainable AI: a discussion with Dan Weld."
48. "Establishment of an Algorithmic Warfare Cross-Functional Team (Project Maven)."
49. "Clocking People's Clocks."
50. Simonite, "Facebook Creates Software That Matches Faces Almost as Well as You Do."
51. Shaud and Lowther, "An Air Force Strategic Vision for 2020–2030," pp. 8–9.
52. Ibid., p. 19.
53. "Unmanned Systems Integrated Roadmap: FY2013-2038," p. 17.
54. Ibid., p. 18.
55. Thompson, Loren, "The Twilight of Network-Centric Warfare."
56. Cockburn, *Kill Chain: the Rise of the High-Tech Assassins*, ch. 1.
57. "AR 15-6 Investigation, 21 February 2010 U.S. Air-to-Ground Engagement in the Vicinity of Shahidi Hassas, Urzugan District, Afghanistan."
58. Crawford, Starr and Hanna, "U.S. general: Human error led to Doctors Without Borders strike."
59. Fisher, "The forgotten story of Iran Air Flight 655."
60. See Singer, *Wired for War*, ch. 6, especially pp. 124–5.
61. Nelson, "Checkers computer becomes invincible."
62. Chinchalkar, "An Upper Bound for the Number of Reachable Positions."
63. Koch, "How the Computer Beat the Go Master."
64. Eadicicco, "This Researcher Programmed the Perfect Poker-Playing Computer."
65. Wilson, "Jeopardy, Schmeopardy."
66. Kelly, "The future of AI? Helping human beings think smarter."
67. Freedburg Jr., "Centaur Army: Bob Work, Robotics, & The Third Offset Strategy."

CONCLUSION ANTICIPATING FUTURE RISKS

1. Smith, "Did ETFs Cause the Flash Crash?"
2. Kirilenko, et al. "The Flash Crash: The Impact of High Frequency Trading on an Electronic Market."
3. Glantz and Kissell, *Multi-Asset Risk Modeling: Techniques for a Global Economy in an Electronic and Algorithmic Trading Era*, p. 258.

Bibliography

ACADEMIC AND MILITARY ARTICLES

Abrahms, Max, "Why Terrorism Does Not Work," *International Security*, vol. 31, no. 2, Fall 2006, pp. 42–78.

Alberts, David S., John J. Garstka, and Frederick P. Stein, *Network Centric Warfare: Developing and Leveraging Information Superiority*, 2nd Edition, February 2000. http://www.dodccrp.org/files/Alberts_NCW.pdf.

Brooks, Rosa, "Drones and the International Rule of Law," *Ethics & International Affairs*, vol. 28, no.1 (2014), pp. 83–103.

Bueno de Mesquita, Ethan, and Eric S. Dickson, "The Propaganda of the Deed: Terrorism, Counterterrorism, and Mobilization," *American Journal of Political Science*, vol. 51, no. 2, April 2007, pp. 364–81.

Byman, Daniel, "Do Targeted Killings Work?," *Foreign Affairs*, Vol. 85, no. 2, March/April 2006, pp. 95–111.

Cebrowski, Arthur K., and John J. Garstka, "Network-Centric Warfare: Its Origin and Future," *United States Naval Institute Proceedings* 124, no. 1, 1998.

Chinchalkar, Shirish, "An Upper Bound for the Number of Reachable Positions," *ICCA Journal*, vol. 19, no. 3, 1996, pp. 181–3.

Cordesman, Anthony, "The Lessons of the Israeli–Lebanon War," *Center for Strategic and International Studies*, March 11, 2008, http://www.csis.org/media/csis/pubs/080311_lessonleb-iswar.pdf.

Crawford, Neta C., "War-related Death, Injury, and Displacement in Afghanistan and Pakistan 2001–2014." *Brown University, Watson Institute: Costs of War*. 2015. http://watson.brown.edu/costsofwar/files/cow/imce/papers/2015/War%20Related%20Casualties%20Afghanistan%20and%20Pakistan%202001–2014%20FIN.pdf.

Cronin, Audrey Kurth, "How al-Qaida Ends: The Decline and Demise of Terrorist Groups," *International Security*, vol. 31, no. 1, Summer 2006, pp. 7–48.

David, Steven R., "Fatal Choices: Israel's Policy of Targeted Killing," *Mideast Security and Policy Studies*, No. 51, September 2002. http://biu.ac.il/Besa/david.pdf.

Dokmanic, Ivan and Reza Parhizkar, Andreas Walther, Yue M. Lu, and Martin Vetterli, "Acoustic echoes reveal room shape," *Proceedings of the National Academy of Sciences of the United States of America*, May 17, 2013. http://www.pnas.org/content/early/2013/06/12/1221464110.full.pdf.

"Drone Strikes Widely Opposed, Global Opinion of Obama Slips, International Policies Faulted," *Pew Global Attitudes Project*, June 13, 2012. http://www.pewglobal.org/2012/06/13/global-opinion-of-obama-slips-international-policies-faulted/.

Eilstrup-Sangiovanni, Mette, and Calvert Jones, "Assessing the Dangers of Illicit Networks: Why al-Qaida May Be Less Threatening Than Many Think," *International Security*, vol. 33, no. 2, Fall 2008, pp. 7–44.

Hafez, Mohammed M. and Joseph M. Hatfield, "Do Targeted Assassinations Work? A Multivariate Analysis of Israeli Counter-Terrorism Effectiveness during Al-Aqsa Uprising," *Studies in Conflict and Terrorism*, Vol. 29, No. 4, June 2006.

Jeffrey, James F., "Why Counterinsurgency Doesn't Work," *Foreign Affairs*, March/April 2015.

Johnston, Patrick B., "Does Decapitation Work? Assessing the Effectiveness of Leadership Targeting in Counterinsurgency Campaigns," *International Security*, Vol. 36, No. 4, Spring 2012, pp. 47–79.

Johnston, Patrick B. and Anoop Sarbahi, "The Impact of US Drone Strikes on Terrorism in Pakistan," April 21, 2015. http://patrickjohnston.info/materials/drones.pdf.

Jordan, Javier, "The Effectiveness of the Drone Campaign against Al Qaeda: A Case Study," *Journal of Strategic Studies*, Vol. 37, No. 1, 2014, pp. 4–29.

Jordan, Jenna, "When Heads Roll: Assessing the Effectiveness of Leadership Decapitation," *Security Studies*, Vol. 18, No. 4, October–December 2009, pp. 4–29.

Kirilenko, Andrei, Albert S. Kyle, Mehrdad Samadi, and Tugkan Tuzun, "The Flash Crash: The Impact of High Frequency Trading on an Electronic Market," October 1, 2010, May 5, 2014. http://www.cftc.gov/idc/groups/public/@economicanalysis/documents/file/oce_flashcrash0314.pdf.

Kissinger, Henry, "The Vietnam Negotiations," *Foreign Affairs*, Vol. 48, No. 2 (January 1969).

Lyall, Jason, and Isaiah Wilson III, "Rage against the Machines: Explaining Outcomes in Counterinsurgency Wars," *International Organization*, Vol. 63, Issue 1, 2009, pp. 67–106.

Mack, Andrew, "Why Big Nations Lose Small Wars: The Politics of Asymmetric Conflict," *World Politics*, vol. 27, no. 2, January 1975, pp. 175–200.

Meigs, General Montgomery C., "Unorthodox Thoughts about Asymmetric Warfare," *Parameters*, Summer 2003, pp. 4–18.

Morris, Lieutenant Colonel Michael F., "Al Qaeda as Insurgency," *U.S. Army War College*, 2005. http://www.dtic.mil/cgi-bin/GetTRDoc?AD=ADA434874.

Price, Bryan, C., "Targeting Top Terrorists: How Leadership Decapitation Contributes to Counterterrorism," *International Security*, Vol. 36, No. 4, Spring 2012, pp. 9–46.

"Public Continues to Back U.S. Drone Attacks," *Pew Research Center*, May 28, 2015. http://www.people-press.org/2015/05/28/public-continues-to-back-u-s-drone-attacks/.

Thompson, Loren B., "The Twilight of Network-Centric Warfare," *Defense Professionals*, August 9, 2010. http://www.lexingtoninstitute.org/the-twilight-of-network-centric-warfare/?a=1&c=1171.

Valentino, Benjamin, Paul Huth and Dylan Balch-Lindsay, "Draining the Sea: Mass Killing and Guerrilla Warfare," *International Organization*, vol. 58, no. 2, 2004, pp. 375–407.

Vego, Milan, "The NCW Illusion," *Armed Forces Journal*, January 2007.

Vickery, Scott A., "Operation Inherent Resolve: An Interim Assessment," *The Washington Institute*, January 13, 2015. http://www.washingtoninstitute.org/policy-analysis/view/operation-inherent-resolve-an-interim-assessment.

BOOKS

Arquilla, John and David Ronfeldt. *Networks and Netwars*, Arlington Virginia: RAND, 2001.

Arreguin-Toft, Ivan, *How the Weak Win Wars: A Theory of Asymmetric Conflict*, Cambridge, United Kingdom: Cambridge University Press, 2005.

Bergen, Peter L., *The Longest War*, New York, New York: The Free Press, 2011.

Cockburn, Andrew, *Kill Chain: the Rise of the High-Tech Assassins*, New York, New York: Henry Holt and Company, 2015.

Coll, Steve, *Ghost Wars*, New York, New York: Penguin Books, 2004.

Crenshaw, Martha ed., *Terrorism in Context*, University Park, Pennsylvania: The Pennsylvania State University Press, fourth printing, 2007.

Galula, David, *Counterinsurgency Warfare: Theory and Practice*, London, United Kingdom: Praeger Security International, 1964, 2006.

——, *Pacification in Algeria*, Arlington, Virginia: Rand, 2006.

Giap, Vo Nguyen, *People's War People's Army*, Honolulu, Hawaii: University Press of the Pacific, 1961.

Glantz, Morton, and Robert Kissell, *Multi-Asset Risk Modeling: Techniques for a Global Economy in an Electronic and Algorithmic Trading Era*, San Diego, California: Academic Press, 2014.

Guevara, Che, *Guerrilla Warfare*, Lincoln, Nebraska: University of Nebraska Press, 1960.

Hoffman, Bruce, *Inside Terrorism*, New York, New York: Columbia University Press, revised and expanded edition, 2006.

Keck, Margaret E., and Kathryn Sikkink, *Activists Beyond Borders*, Ithaca, New York: Cornell University Press, 1998.

Levitt, Matthew, *Hamas: Politics, Charity, and Terrorism in the Service of Jihad*, New Haven, Connecticut: Yale University Press, 2006.

Mao Zedong, "The Struggle in the Chingkang Mountains," 1928.

—— "Problems of Strategy in China's Revolutionary War," 1936.

—— "Problems of Strategy in Guerrilla War Against Japan," 1938.

—— "On Protracted War," 1938.

—— *Selected Military Writings of Mao Tse-Tung*, Beijing, China: Foreign Language Press, 1968.

Mueller, John, *War, Presidents, and Public Opinion*, Hoboken, New Jersey: John Wiley and Sons Inc., 1973.

Al-Muqrin, 'Abd Al-'Aziz, *A Practical Course for Guerrilla War*, translated by Norman Cigar in *Al-Qa'ida's Doctrine for Insurgency*, Dulles, Virginia: Potomac Books Inc., 2009.

Nacos, Brigitte L., *Mass-Mediated Terrorism*, Lanham, Maryland: Rowman & Littlefield Publishers, Inc., 2007.

Nagl, John A., *Learning to Eat Soup with a Knife: Counterinsurgency Lessons from Malaya and Vietnam*, Chicago, Illinois: University of Chicago Press, 2002, 2005.

Norton, Augustus Richard, *Hezbollah: A Short History*, Princeton, New Jersey: Princeton University Press, 2007.

Schelling, Thomas, *Arms and Influence*, New Haven, Connecticut: Yale University Press, 1966.

Shaud, John A. and Adam B. Lowther, "An Air Force Strategic Vision for 2020–2030," *Strategic Studies Quarterly* (Spring 2011).

Singer, P.W., *Wired for War: The Robotics Revolution and Conflict in the 21st Century*, New York, New York: Penguin Press HC, 2009.

Sonka, Milan, et al., *Image Processing, Analysis, and Machine Vision*, CL Engineering, 2007.

Trinquier, Roger, *Modern Warfare: A French View of Counterinsurgency*, London, United Kingdom: Praeger Security International, 1964, 2006.

Valeriano, Napolean D., and Charles T.R. Bohannan, *Counter-Guerrilla Operations: The Philippine Experience*, London, United Kingdom: Praeger Security International, 1962, 2006.

Wagner, William, *Lightning Bugs and Other Reconnaissance Drones*, Washington, DC: Armed Forces Journal, 1982.

Weimann, Gabriel, *Terror on the internet*, United States Institute for Peace: Washington, DC, 2006.

DATA COLLECTIONS

"Afghanistan," *Gallup*. http://www.gallup.com/poll/116233/afghanistan.aspx.

"Covert Drone War," *The Bureau of Investigative Journalism*. https://www.thebureauinvestigates.com/category/projects/drones/.

Dobrydney, David, "Combined Forces Air Component Command Airpower Statistics," *United States Air Force Central Command*, January 6, 2013. http://www.afcent.af.mil/shared/media/document/AFD-130106-001.pdf.

"Drone Wars," *New America Foundation*. http://securitydata.newamerica.net/about.html.

"Fatalities in Terrorist Violence in Pakistan 2003–2015," *South Asia Terrorism Portal*, December 20, 2015. http://www.satp.org/satporgtp/countries/pakistan/database/casualties.htm.

"Obama's Covert War in Yemen," *New American Foundation*, http://yemendrones.newamerica.net/.

"Operation Enduring Freedom," *iCasualties.org*. http://icasualties.org/oef/.

"Operation Iraqi Freedom," *iCasualies.org*. http://icasualties.org/iraq/index.aspx.

"SIPRI (Stockholm International Peace Research Institute) Military Expenditure Database." http://milexdata.sipri.org/.

"Terrorism Against Israel: Comprehensive Listing of Fatalities," *Jewish Virtual Library*. http://www.jewishvirtuallibrary.org/jsource/Terrorism/victims.html#1993.

"Terrorism deaths in Israel – 1920–1999," *Israel Ministry of Foreign Affairs*. http://www.mfa.gov.il/MFA/MFA-Archive/2000/Pages/Terrorism%20deaths%20in%20Israel%20-%201920–1999.aspx.

"UCDP Conflict Encyclopedia," *UCDP Database*. http://www.ucdp.uu.se/gpdatabase/search.php.

Wayner, Pete "American Approval Rating (Percent) of War in Afghanistan," http://www-958.ibm.com/software/data/cognos/manyeyes/datasets/american-approval-rating-percent-o/versions/1.

"World of Drones: Military," *New America Foundation*. http://securitydata.newamerica.net/world-drones.html.

"The Year of the Drone: An Analysis of U.S. Drone Strikes in Pakistan, 2004–2012," *New America Foundation*, http://counterterrorism.newamerica.net/drones.

GOVERNMENTS, INTERNATIONAL AGENCIES, NGOS AND RELATED ORGANIZATIONS

"2011 Request for Information on Tamerlan Tsarnaev from Foreign Government," *FBI*, April 19, 2013. https://www.fbi.gov/news/pressrel/press-releases/2011-request-for-information-on-tamerlan-tsarnaev-from-foreign-government.

"The 9/11 Commission Report," *National Commission on Terrorist Attacks upon the United States*, New York, New York: W.W. Norton & Company, 2006.

"Afghanistan: Annual Report 2012 Protection of Civilians in Armed Combat," *United Nations Assistance Mission in Afghanistan and UN Office of the High Commissioner for Human Rights*, February, 2013. http://unama.unmissions.org/LinkClick.aspx?fileticket=K0B5RL2XYcU%3d&tabid=12254&language=en-US.

"The Al-Qaida Papers – Drones," *Associated Press*, June 17, 2011. http://hosted. ap.org/specials/interactives/_international/_pdfs/al-qaida-papers-drones. pdf.

"Al-Qaida Sanctions List," *Security Council Committee pursuant to resolutions 1267 (1999) and 1989 (2011) concerning Al-Qaida and associated individuals and entities*, July 11, 2013.

"Analysis of the Fiscal Year 2012 Pentagon Spending Request," *CostofWar.com*, February 15, 2011. http://costofwar.com/en/publications/2011/analysis-fiscal-year-2012-pentagon-spending-request/ http://www.un.org/sc/ committees/1267/AQList.htm#alqaedaent.

"AR 15–6 Investigation, 21 February 2010 U.S. Air-to-Ground Engagement in the Vicinity of Shahidi Hassas, Urzugan District, Afghanistan," May 2, 2010. https://www.aclu.org/drone-foia-department-defense-uruzgan-investigation-documents.

"Armed Robots March into Battle," *United States Department of Defense*, December 6, 2004. http://www.defense.gov/transformation/articles/ 2004–12/ta120604c.html.

"Attacks on Israeli civilians by Palestinians," *B'Tselem*, July 24, 2014. http:// www.btselem.org/israeli_civilians/qassam_missiles#data.

"A Wedding That Became a Funeral," *Human Rights Watch*, February 19, 2014. https://www.hrw.org/report/2014/02/19/wedding-became-funeral/us-drone-attack-marriage-procession-yemen.

Cacace, Gary, "Affidavit of Special Agent Gary S. Cacace: 11-mj-4270-tsh" published by *intelwire*. http://intelwire.egoplex.com/Ferdaus-Affidavit.pdf.

"Camera Restrictions in New York," *411 New York*. http://411newyork.org/ guide/2008/09/11/camera-restrictions-in-new-york/.

"Causal Factors & Risk Ratings," *UK Airprox Board*. http://www.airproxboard. org.uk/default.aspx?catid=423&pagetype=90&pageid=5637.

"Counterinsurgency Field Manual," *U.S. Army and Marine Corps*, Chicago, Illinois: University of Chicago Press, 2007.

"Defense Budget Priorities and Choices," *U.S. Department of Defense*, January 2012. www.defense.gov/news/Defense_Budget_Priorities.pdf.

"Designated Foreign Terrorist Organizations," *U.S. Department of State*, September 28, 2012. http://www.state.gov/j/ct/rls/other/des/123085. htm.

Dobbing, Mary and Chris Cole, "Israel and the Drone Wars," *Drone Wars UK*, January 2014. https://dronewarsuk.files.wordpress.com/2014/01/israel-and-the-drone-wars.pdf.

"Enduring Strategic Partnership Agreement between the Islamic Republic of Afghanistan and the United States of America," May 2, 2012. http://www. scribd.com/doc/92057506/Afghan-US-Strategic-Pact-Full-Text.

"Establishment of an Algorithmic Warfare Cross-Functional Team (Project Maven)," *Office of the Deputy Secretary of Defense*, April 26, 2017. https:// www.govexec.com/media/gbc/docs/pdfs_edit/establishment_of_the_ awcft_project_maven.pdf.

"Final Vote Results for Roll Call 342," *Clerk of the United States House of Representatives*, September 14, 2001. http://clerk.house.gov/evs/2001/roll342.xml.

"Fiscal Year 2016 Budget Request," *United States Department of Defense*, February 2015. http://comptroller.defense.gov/Portals/45/Documents/defbudget/fy2016/FY2016_Budget_Request_Overview_Book.pdf.

"Foreign Assistance: Summary Tables Fiscal Year 2015," *United States Department of State*, 2014. http://www.state.gov/documents/organization/224071.pdf.

"Foreign Fighters: An Updated Assessment of the Flow of Foreign Fighters into Syria and Iraq," *The Soufan Group*, December 2015. http://soufangroup.com/wp-content/uploads/2015/12/TSG_ForeignFightersUpdate3.pdf.

Gordon, Philip H., "Madrid Bombing and U.S. Policy," *Brookings*, March 31, 2004. http://www.brookings.edu/research/testimony/2004/03/31europe-gordon.

"Hamas' Weapons Arsenal Continues to Grow," *IDF Blog*, February 14, 2012. http://www.idfblog.com/hamas/2012/02/14/hamas-weapons-arsenal-continues-grow/.

"HM announces measures to enhance security," *Press Information Bureau, Government of India*, December 11, 2008. http://pib.nic.in/newsite/erelease.aspx?relid=45446.

"The Implementation of Network-Centric Warfare," *The Office of Force Transformation*, 2005. http://www.au.af.mil/au/awc/awcgate/transformation/oft_implementation_ncw.pdf.

"It's (a) Grand! FAA Passes 1,000 UAS Section 333 Exemptions," *Federal Aviation Administration*, August 4, 2015. https://www.faa.gov/news/updates/?newsId=83395.

"Joint Doctrine for Information Operations" (Joint Pub 3–13), October 1998. http://www.c4i.org/jp3_13.pdf.

"Letter from al-Zawahiri to al-Zarqawi," *Globalsecurity.org*, July 9, 2005. http://www.globalsecurity.org/security/library/report/2005/zawahiri-zarqawi letter_9jul2005.htm.

"Miniature surveillance helicopters help protect front line troops," *gov.uk*, February 4, 2013. https://www.gov.uk/government/news/miniature-surveillance-helicopters-help-protect-front-line-troops.

"Operation TELIC: British Casualties and Fatalities," *National Archives UK*. http://webarchive.nationalarchives.gov.uk/+/http://www.operations.mod.uk/telic/casualties.htm.

"Part 8 – Casualty Handling," *Report of the DoD Commission on Beirut International Airport Terrorist Act*, October 23, 1983. http://www.ibiblio.org/hyperwar/AMH/XX/MidEast/Lebanon-1982–1984/DOD-Report/Beirut-8.html.

Petraeus, David H., "Report to Congress on the Situation in Iraq," September 10, 2007. http://www.defense.gov/pubs/pdfs/Petraeus-Testimony2007 0910.pdf.

"Press Release – DOT and FAA Propose New Rules for Small Unmanned Aircraft Systems," *Federal Aviation Administration*, February 15, 2015. http://www.faa.gov/news/press_releases/news_story.cfm?newsId=18295.

"Press Release – FAA Announces Small UAS Registration Rule," *Federal Aviation Administration*, December 14, 2015. http://www.faa.gov/news/press_releases/news_story.cfm?newsId=19856.

"Press Release – New FAA Rules for Small Unmanned Aircraft Systems Go Into Effect," *Federal Aviation Administration*, August 29, 2016. https://www.faa.gov/news/press_releases/news_story.cfm?newsId=20734.

"Proscribed Terrorist Organizations," *United Kingdom Home Office*, March 27, 2015. https://www.gov.uk/government/uploads/system/uploads/attachment_data/file/472956/Proscription-update-20151030.pdf.

"'Reaper' moniker given to MQ-9 unmanned aerial vehicle," *U.S. Air Force*, September 9, 2006. http://www.af.mil/news/story.asp?storyID=123027012.

"Registration and Marking Requirements for Small Unmanned Aircraft: A Rule by the Federal Aviation Administration," *Federal Register*, December 16, 2015. https://www.federalregister.gov/articles/2015/12/16/2015–31750/registration-and-marking-requirements-for-small-unmanned-aircraft#h-52.

"Reports and Analysis," *UK Airprox Board*. http://www.airproxboard.org.uk/default.aspx?catid=423&pagetype=90&pageid=5638.

Roudik, Peter, "Russia: Commercial and Private Drones to Be Outlawed in Moscow," *The Law Library of Congress*, June 25, 2015. http://www.loc.gov/law/foreign-news/article/russia-commercial-and-private-drones-to-be-outlawed-in-moscow/.

"SUA Operators," *Civil Aviation Authority*, December 21, 2015. https://www.caa.co.uk/uploadedFiles/CAA/Content/Standard_Content/Commercial_industry/Aircraft/Unmanned_aircraft/CAAApprovedSUAOperators21122015.pdf.

"Text of Authorization for Use of Military Force," *govtrack.us*, September 18, 2001. https://www.govtrack.us/congress/bills/107/sjres23/text.

"UAS Sightings Reports," *Federal Aviation Administration*, March 25, 2016. http://www.faa.gov/uas/resources/uas_sightings_report/.

"UK Drone Strike Stats," *Drone Wars UK*, November 16, 2015. http://dronewars.net/uk-drone-strike-list-2/.

"UK forces: operations in Afghanistan," *gov.uk*, June 18, 2013. https://www.gov.uk/uk-forces-operations-in-afghanistan.

"Unmanned Systems Integrated Roadmap: FY2013–2038," *United States Department of Defense*, December 23, 2013.

"The Warsaw Ghetto Uprising," *The United States Holocaust Museum*. http://www.ushmm.org/outreach/wgupris.htm.

"What is the FATF?," *Financial Action Task Force*. http://www.fatf-gafi.org/pages/aboutus/.

"Winograd Committee submits final report," *Israel Ministry of Foreign Affairs*, January 30, 2008. http://www.mfa.gov.il/mfa/mfa-archive/2008/pages/winograd%20committee%20submits%20final%20report%2030-jan-2008.aspx.

"Yemen's al-Qaeda: Expanding the Base," *International Crisis Group*, February 2, 2017. https://www.crisisgroup.org/middle-east-north-africa/gulf-and-arabian-peninsula/yemen/174-yemen-s-al-qaeda-expanding-base.

NEWS MEDIA

"Airport Incident 'Was Terrorism,'" *BBC News*, July 1, 2007. http://news.bbc.co.uk/2/hi/uk_news/scotland/6257846.stm.

Allam, Hannah, "Is imam a terror recruiter or just an incendiary preacher?," *McClatchy*, November 20, 2009. http://www.mcclatchydc.com/news/nation-world/world/article24564601.html.

Allen, Nick, "WikiLeaks: Yemen Covered Up US Drone Strikes," *Telegraph*, November 28, 2010. http://www.telegraph.co.uk/news/worldnews/middleeast/yemen/8166610/WikiLeaks-Yemen-covered-up-US-drone-strikes.html.

Almosawa, Shuaib and Rod Nordland, "U.S. Fears Chaos as Government of Yemen Falls," *New York Times*, January 22, 2015. http://www.nytimes.com/2015/01/23/world/middleeast/yemen-houthi-crisis-sana.html.

"Amazon Unveils Futuristic Plan: Delivery by Drone," *CBS News*, December 1, 2013. http://www.cbsnews.com/news/amazon-unveils-futuristic-plan-delivery-by-drone/.

"Ambush in Mogadishu," *PBS Frontline*, http://www.pbs.org/wgbh/pages/frontline/shows/ambush.

Andrews, Todd, "A commercial airline collided with a drone in Canada, a first in North America," *Washington Post*, October 16, 2017. https://www.washingtonpost.com/news/morning-mix/wp/2017/10/16/a-commercial-airplane-collided-with-a-drone-in-canada-a-first-in-north-america/?tid=hybrid_collaborative_1_na&utm_term=.52b059a93447.

Andrews, Wilson and Todd Lindeman, "The Black Budget," *Washington Post*, August 29, 2013. http://www.washingtonpost.com/wp-srv/special/national/black-budget/.

"AP–GfK Poll: Americans approve of drone strikes on terrorists," *AP-GfK*, May 1, 2015. http://ap-gfkpoll.com/featured/findings-from-our-latest-poll-16.

Atkinson, Rick, "Night of a Thousand Casualties; Battle Triggered U.S. Decision to Withdraw From Somalia Series," *Washington Post*, January 1994.

——, "Left of Boom: The Struggle to Defeat Roadside Bombs," *Washington Post*, October 2007, http://www.washingtonpost.com/wp-srv/world/specials/leftofboom/index.html.

Axe, David, "Predator Drones Once Shot Back at Jets … But Sucked At It," *Wired*, November 9, 2012. http://www.wired.com/dangerroom/2012/11/predator-defenseless/

Baker, Aryn, "Why Al-Qaeda Kicked Out Its Deadly Syrian Franchise," *Time*, February 3, 2014. http://time.com/3469/why-al-qaeda-kicked-out-its-deadly-syria-franchise/.

Barnett, David, "Hezbollah takes responsibility for last week's drone over Israel," *Long War Journal*, October 11, 2012. http://www.longwarjournal.org/archives/2012/10/netanyahu_hezbollah.php.

Barzak, Ibrahim, "After attack on jeep, Israeli army kills 4 in Gaza," *Associated Press*, November 10, 2012. http://news.yahoo.com/attack-jeep-israeli-army-kills-4-gaza-175914332.html.

Becker, Jo and Scott Shane, "Secret 'Kill List' Proves a Test of Obama's Principles and Will," *New York Times*, May 29, 2012. http://www.nytimes.com/2012/05/29/world/obamas-leadership-in-war-on-al-qaeda.html.

Becker, Olivia, "ISIS Has a Really Slick and Sophisticated Media Department," *Vice News*, July 12, 2014. https://news.vice.com/article/isis-has-a-really-slick-and-sophisticated-media-department.

"Beirut Barracks Attack Remembered," *CBS News*, October 23, 2003, http://www.cbsnews.com/stories/2003/10/23/world/main579638.shtml.

Beizer, Doug, "BAE to Develop Surveillance System," *Washington Post*, November 12, 2007. http://www.washingtonpost.com/wp-dyn/content/article/2007/11/11/AR2007111101348.html.

Belfiore, Michael, "Carnegie Takes First in DARPA's Urban Challenge," *Wired*, November 4, 2007. http://www.wired.com/dangerroom/2007/11/darpa-names-win/.

Benson, Pam, "Is the core of al Qaeda on its last legs?," *CNN*, April 27, 2012. http://security.blogs.cnn.com/2012/04/27/is-the-core-of-al-qaeda-on-its-last-legs/.

Bergen, Peter, "Time to declare victory: al Qaeda is defeated," *CNN*, June 27, 2012. http://security.blogs.cnn.com/2012/06/27/time-to-declare-victory-al-qaeda-is-defeated-opinion/.

——, "Why does ISIS keep making enemies?," *CNN*, February 18, 2015. http://www.cnn.com/2015/02/16/opinion/bergen-isis-enemies/.

——, "Secrets of the bin Laden treasure-trove," *CNN*, May 20, 2015. http://www.cnn.com/2015/05/20/opinions/bergen-bin-laden-document-trove/.

Bergen, Peter L. and Jennifer Rowland, "Obama Ramps Up Covert War in Yemen," *New America Foundation*, June 11, 2012. http://www.newamerica.net/publications/articles/2012/obama_ramps_up_covert_war_in_yemen_68427.

Berger, J.M., "How ISIS Games Twitter," *The Atlantic*, June 16, 2014. http://www.theatlantic.com/international/archive/2014/06/isis-iraq-twitter-social-media-strategy/372856/.

——, "The Islamic State vs. al Qaeda," *Foreign Policy*, September 2, 2014. http://foreignpolicy.com/2014/09/02/the-islamic-state-vs-al-qaeda/.

Bergman, Ronen, "Hezbollah boosting drone unit," *Ynet News*, April 27, 2012. http://www.ynetnews.com/articles/0,7340,L-4221414,00.html.

Bialik, Carl, "Shadowy Figure: Al Qaeda's Size is Hard to Measure," *Wall Street Journal*, September 10, 2011. http://www.wsj.com/articles/SB10001424053 111903285704576560593124523206.

Bidgood, Jess, "Massachusetts Man Gets 17 Years for Terrorist Plot," *New York Times*, November 2, 2012. http://www.nytimes.com/2012/11/02/us/rezwan-ferdaus-of-massachusetts-gets-17-years-in-terrorist-plot.html?_r=0.

Bittel, Jason, "Studying Hurricanes With Swarms of Smart Drones," *Slate*, June 7, 2013. http://www.slate.com/blogs/future_tense/2013/06/07/hurricane_research_drones_small_autonomous_submarine_and_plane_are_future.html.

Booth, Robert and Ian Black, "WikiLeaks cables: Yemen offered US 'open door' to attack al-Qaida on its soil," *Guardian*, December 3, 2010. http://www.theguardian.com/world/2010/dec/03/wikileaks-yemen-us-attack-al-qaida.

Botelho, Greg and Barbara Starr, "U.S. 'reasonably certain' drone strike killed ISIS mouthpiece 'Jihadi John,'" *CNN*, November 14, 2015. http://www.cnn.com/2015/11/13/middleeast/jihadi-john-airstrike-target/.

Bowden, Mark, "A defining battle," *Philadelphia Inquirer*, November 16, 1997. http://inquirer.philly.com/packages/somalia/nov16/rang16.asp.

Brandom, Russell, "How Brazil is trying (and failing) to keep drones away from the Olympics," *The Verge*, August 8, 2016. http://www.theverge.com/2016/8/8/12402972/olympics-rio-2016-anti-drone-jamming-public-safety.

"Britain's jihadi kill list," *The Economist*, September 12, 2015. http://www.economist.com/news/middle-east-and-africa/21664154-killing-drone-strike-british-jihadist-syria-will-not-be.

Byman, Daniel, "Can Al Qaeda in the Arabian Peninsula Survive the Death of Its Leader?," *Foreign Policy*, June 16, 2015. http://foreignpolicy.com/2015/06/16/yemen-al-qaeda-zawahiri-wuhayshi/.

Chung, Emily, "Amazon tests delivery drones at a secret site in Canada—here's why," *CBC News*, March 30, 2014. http://www.cbc.ca/news/technology/amazon-tests-delivery-drones-at-a-secret-site-in-canada-here-s-why-1.3015425.

"CIA 'killed al Qaeda suspects' in Yemen," *BBC News*, November 5, 2002. http://news.bbc.co.uk/2/hi/2402479.stm.

"Clocking People's Clocks," *The Economist*, August 23, 2014. http://www.economist.com/news/science-and-technology/21613160-facial-recognition-systems-are-getting-better-clocking-peoples-clocks.

Cobain, Ian, "London bombings: the day the anti-terrorism rules changed," *Guardian*, July 7, 2010. http://www.theguardian.com/uk/2010/jul/07/london-bombings-anti-terrorism.

Cohen, Tom, "When can a government kill its own people?," *CNN*, February 11, 2014. http://www.cnn.com/2014/02/10/politics/us-killing-americans/.

Cole, Juan, "Al-Awlaqi Should have been Tried in Absentia," *Informed Comment*, October 1, 2011. http://www.juancole.com/2011/10/al-awlaqi-should-have-been-tried-in-absentia.html.

"Could The Use Of Flying Death Robots Be Hurting America's Reputation Worldwide?," *The Onion*. http://www.theonion.com/video/could-the-use-of-flying-death-robots-be-hurting-am-27601.

Crawford, Jamie, Barbara Starr and Jason Hanna, "U.S. general: Human error led to Doctors Without Borders strike," *CNN*, November 25, 2015. http://www.cnn.com/2015/11/25/politics/afghanistan-kunduz-doctors-without-borders-hospital/.

Crowcroft, Orlando, "Al-Qaeda seizes major airport and Mukalla oil terminal in Southern Yemen," *International Business Times*, April 16, 2015. http://www.ibtimes.co.uk/al-qaeda-seizes-major-airport-mukalla-oil-terminal-southern-yemen-1496805.

——, "Yemen: Who is the new leader of al-Qaeda in the Arabian Peninsula Qasim al-Raymi?," *International Business Times*, June 16, 2015. http://www.ibtimes.co.uk/yemen-who-new-leader-al-qaeda-arabian-peninsula-qasim-al-raymi-1506432.

Cruickshank, Paul and Tim Lister, "Al Qaeda in the Arabian Peninsula confirms links to underwear bomber," *CNN*, December 23, 2014. http://www.cnn.com/2014/12/23/justice/al-qaeda-underwear-bomber/.

Danigelis, Alyssa, "Tiny Dragonfly UAV Flies and Hovers to Spy," *Discovery News*, November 8, 2012. http://news.discovery.com/tech/dragonfly-uav-121108.htm.

Davies, Alex, "The Marines' Self-Flying Chopper Survives a Three-Year Tour," *Wired*, July 30, 2014. http://www.wired.com/2014/07/kmax-autonomous-helicopter/.

"Dead, captured and wanted," *CNN*, April 27, 2012. http://security.blogs.cnn.com/2012/04/27/dead-captured-and-wanted-2/?hpt=hp_c1.

Dehghan, Saeed Kamali, "Iran supplied Hamas with Fajr-5 missile technology," *Guardian*, November 21, 2012. http://www.guardian.co.uk/world/2012/nov/21/iran-supplied-hamas-missile-technology.

DeYoung, Karen and Greg Miller, "White House releases its count of civilian deaths in counterterrorism operations under Obama," *Washington Post*, July 1, 2016. https://www.washingtonpost.com/world/national-security/white-house-releases-its-count-of-civilian-deaths-in-counterterrorism-operations-under-obama/2016/07/01/3196aa1e-3fa2–11e6–80bc-d06711f d2125_story.html.

Diamond, Jeremy, "U.S. drone strike accidentally killed 2 hostages," *CNN*, April 23, 2015. http://www.cnn.com/2015/04/23/politics/white-house-hostages-killed/.

"Dogfight between MQ-1 Predator drone and MiG-25 Foxbat," *CBS News*. http://www.youtube.com/watch?v=wWUR3sgKUV8.

Donohue, Pete and Dave Goldiner, "How often to birds cause plane crashes?", *New York Daily News*, January 16, 2009. http://www.nydailynews.com/new-york/birds-plane-crashes-article-1.361189.

Dorell, Oren, "Iranian support for Yemen's Houthis goes back years," *USA Today*, April 20, 2015. http://www.usatoday.com/story/news/world/2015/04/20/iran-support-for-yemen-houthis-goes-back-years/26095101/.

Dozier, Kimberly, "Report: U.S. drone strike may have killed up to a dozen civilians in Yemen," *Associated Press*, February 20, 2014. http://www.pbs.org/newshour/rundown/report-u-s-drones-may-killed-civilians/.

"'Drone' hits British Airways plane approaching Heathrow Airport," *BBC News*, April 17, 2016. http://www.bbc.com/news/uk-36067591.

"Drone Shot Down Over Iran 'Lost' Over Afghanistan Last Week," *Telegraph*, December 4, 2001. http://www.telegraph.co.uk/news/worldnews/middleeast/iran/8934451/Drone-shot-down-over-Iran-lost-over-Afghanistan-last-week.html.

Dunlap, Charles, "We Still Need the Big Guns," *New York Times*, January 9, 2008.

Eadicicco, Lisa, "This Researcher Programmed the Perfect Poker-Playing Computer," *Time*, February 1, 2017. http://time.com/4656011/artificial-intelligence-ai-poker-tournament-libratus-cmu/.

Ephron, Dan, "Hizbullah's Worrisome Weapon," *Newsweek*, September 11, 2006.

Fadel, Leith, "ISIS Drone Downed by the Syrian Army at Kuweires Airbase in Aleppo," *Al Masdar News*, May 15, 2015. http://www.almasdarnews.com/article/isis-drone-downed-by-the-syrian-army-at-kuweires-airbase-in-aleppo/.

Ferran, Lee, "Top US Spy: Intel Cooperation with Pakistan 'On the Upswing,'" *ABC News*, April 14, 2014. http://abcnews.go.com/blogs/headlines/2014/04/top-us-spy-intel-coopration-with-pakistan-on-the-upswing/.

Fishel, Jusin, Pierre Thomas, Mike Levine and Jack Date, "EXCLUSIVE: Undercover DHS Tests Find Security Failures at US Airports," *ABC News*, June 1, 2015. http://abcnews.go.com/US/exclusive-undercover-dhs-tests-find-widespread-security-failures/story?id=31434881.

Fisher, Max, "The forgotten story of Iran Air Flight 655," *Washington Post*, October 16, 2013. http://www.washingtonpost.com/blogs/worldviews/wp/2013/10/16/the-forgotten-story-of-iran-air-flight-655/.

Fishman, Alex, "Uncovering the missiles," *Ynet News*, March 16, 2011. http://www.ynetnews.com/articles/0,7340,L-4043392,00.html.

Friedersdorf, Conor, "Yes, Pakistanis Really Do Hate America's Killer Drones," *The Atlantic*, January 24, 2013. http://www.theatlantic.com/international/archive/2013/01/yes-pakistanis-really-do-hate-americas-killer-drones/272468/.

Gallagher, Ryan, "Could the Pentagon's 1.8 Gigapixel Drone Camera By Used for Domestic Surveillance?," *Slate*, February 6, 2013. http://www.slate.com/blogs/future_tense/2013/02/06/argus_is_could_the_pentagon_s_1_8_gigapixel_drone_camera_be_used_for_domestic.html.

Gallagher, Sean, "France cries foul at World Cup "spy drone," *ARS Technica*, June 16, 2014. http://arstechnica.com/information-technology/2014/06/france-cries-foul-at-world-cup-spy-drone/.

Gannon, Kathy and Sebastian Abbot, "Criticism alters US drone program in Pakistan," *Associated Press*, July 25, 2013. http://www.stripes.com/news/middle-east/criticism-alters-us-drone-program-in-pakistan-1.232281.

Gellman, Barton and Laura Poitras, "U.S., British intelligence mining data from nine U.S. Internet companies in broad secret program," *Washington Post*, June 7, 2013.

Geoghegan, Tom, "Innocent photographer or terrorist?," *BBC News*, April 17, 2008. http://news.bbc.co.uk/2/hi/technology/7351252.stm.

Gorman, Siobhan and Yochi J. Dreazen and August Cole, "Insurgents Hack U.S. Drones," *Wall Street Journal*, December 17, 2009. http://online.wsj.com/article/SB126102247889095011.html.

Greenberg, Andy, "Flying Drone Can Crack Wi-Fi Networks, Snoop on Cell Phones," *Forbes*, July 28, 2011. http://www.forbes.com/sites/andygreenberg/2011/07/28/flying-drone-can-crack-wifi-networks-snoop-on-cell-phones/.

Greenfield, Danya, "The Case Against Drone Strikes on People Who Only 'Act' Like Terrorists," *The Atlantic*, August 19, 2013. http://www.theatlantic.com/international/archive/2013/08/the-case-against-drone-strikes-on-people-who-only-act-like-terrorists/278744/.

Grey, Stephen, "U.S. Kills al Qaeda Leaders by Remote Control," *Sunday Times*, November 18, 2001. http://www.freerepublic.com/focus/f-news/573934/posts.

Guerin, Orla, "US drone war in Pakistan prompts fear and anger," *BBC News*, October 5, 2012. http://www.bbc.com/news/world-asia-19842410.

Hall, John, "ISIS propaganda, Call of Duty Style," *Daily Mail*, December 12, 2014. http://www.dailymail.co.uk/news/article-2871389/ISIS-propaganda-Call-Of-Duty-style-Latest-footage-shows-drone-s-view-battle-ravaged-streets-Kobane-swooping-gun-battles-ground.html.

Hanna, Jason, Michael Martinez and Jennifer Deaton, "ISIS publishes photo of what it says is bomb that downed Russian plane," *CNN*, November 19, 2015. http://www.cnn.com/2015/11/18/middleeast/metrojet-crash-dabiq-claim/.

Harel, Amos, "Air Force: Hezbollah drone flew over Israel for five minutes," *Haaretz*, November 9, 2004. http://www.haaretz.com/print-edition/news/air-force-hezbollah-drone-flew-over-israel-for-five-minutes-1.139744.

Harel, Amos et al., "Hezbollah drone brought down over Galilee held 30 kg of explosives," *Haaretz*, August 14, 2006. http://www.haaretz.com/news/hezbollah-drone-brought-down-over-galilee-held-30-kg-of-explosives-1.195115.

Hennigan, W.J., "Pentagon to soon deploy pint-sized but lethal Switchblade drones," *Los Angeles Times*, June 11, 2012. http://articles.latimes.com/2012/jun/11/business/la-fi-kamikaze-drone-20120611.

——, "United Arab Emirates set to buy U.S. Predator drones," *Los Angeles Times*, February 22, 2013. http://articles.latimes.com/2013/feb/22/business/la-fi-predator-drone-sale-20130223.

Herbert, Keith, "Boston Marathon timeline: from attack to capture," *Newsday*, April 20, 2013. http://www.newsday.com/news/nation/boston-marathon-timeline-from-attack-to-capture-1.5112336.

"Hezbollah Drones Target Al-Nusra Front's Positions at Syrian Border," *FARS*, September 21, 2014. http://en.farsnews.com/newstext.aspx?nn= 13930630001247.

"Hezbollah's Rocket Force," *BBC News*, July 18, 2006. http://news.bbc.co.uk/2/hi/middle_east/5187974.stm.

"Hezbollah strike Qalamoun jihadist with help of drone," *Daily Star*, May 27, 2015. http://www.dailystar.com.lb/News/Lebanon-News/2015/May-27/299419-hezbollah-annihilates-nusra-unit-in-qalamoun-al-manar.ashx.

"Hospital staff stunned as doctors are questioned," *Guardian*, July 2, 2007.

"IAEA: Syria tried to build nuclear reactor," *Associated Press*, April 28, 2011. http://www.ynetnews.com/articles/0,7340,L-4062001,00.html.

"Islamic State to halve fighters' salaries as cost of waging terror starts to bite," *Guardian*, January 19, 2016. https://www.theguardian.com/world/2016/jan/20/islamic-state-to-halve-fighters-salaries-as-cost-of-waging-terror-starts-to-bite.

"ISIS drone shot down by Peshmerga," *Rudaw*, May 5, 2015. http://rudaw.net/english/kurdistan/05052015.

"Islamic State and the crisis in Iraq and Syria in maps," *BBC*, November 3, 2017. http://www.bbc.com/news/world-middle-east-27838034.

"Israel deploys 'Iron Dome' rocket shield," *Al Jazeera*, March 27, 2011. http://www.aljazeera.com/news/middleeast/2011/03/201132718224159699.html.

Issacharoff, Avi, "Israel raises Hezbollah rocket estimate to 150,000," *Times of Israel*, November 12, 2015. http://www.timesofisrael.com/israel-raises-hezbollah-rocket-estimate-to-150000/.

Janik, Rachel and Mitchell Armentrout, "Industry looks to use drones for commercial purposes," *Miami Herald*, April 29, 2013. http://www.miamiherald.com/2013/04/29/3371170/industry-looks-to-use-drones-for.html.

Johnson, Kevin, "Man accused of plotting drone attacks on Pentagon, Capital," *USA Today*, September 29, 2011. http://usatoday30.usatoday.com/news/washington/story/2011-09-28/DC-terrorist-plot-drone/50593792/1.

Katz, Yaakov, "IAF deploys third Iron Dome battery outside Ashdod," *Jerusalem Post*, August 31, 2011. http://www.jpost.com/Defense/IAF-deploys-third-Iron-Dome-battery-outside-Ashdod.

Katz, Yaakov and Yaakov Lappin, "Iron Dome ups its interception rate to over 90%," *Jerusalem Post*, March 10, 2012. http://www.jpost.com/Defense/Iron-Dome-ups-its-interception-rate-to-over-90-percent.

Khan, Behroz, "Pakistan Taliban: US Drone Strikes Forcing Militants Underground," *Christian Science Monitor*, March 15, 2010. http://www.csmonitor.com/World/Asia-South-Central/2010/0315/Pakistan-Taliban-US-drone-strikes-forcing-militants-underground.

Kharpal, Arjun, "Amazon wins patent for a flying warehouse that will deploy drones to deliver parcels in minutes," *CNBC*, December 30, 2016. http://www.cnbc.com/2016/12/29/amazon-flying-warehouse-deploy-delivery-drones-patent.html.

Khatchadourian, Raffi, "Azzam the American," *New Yorker*, January 22, 2007. http://www.newyorker.com/magazine/2007/01/22/azzam-the-american.

Khoury, Hala, "Last French peacekeepers ready to leave Beirut," *United Press International*, March 31, 1984.

Knausgaard, Karl Ove, "The Inexplicable: Inside the mind of a killer," *New Yorker*, May 25, 2015. http://www.newyorker.com/magazine/2015/05/25/the-inexplicable.

Koch, Christof, "How the Computer Beat the Go Master," *Scientific American*, March 19, 2016. https://www.scientificamerican.com/article/how-the-computer-beat-the-go-master/.

Koebler, Jason, "Burrito Bomber Attacks Hunger with Drone-Delivered Mexican Food," *U.S. News and World Report*, December 21, 2012. http://www.usnews.com/news/articles/2012/12/21/burrito-bomber-starts-the-drone-delivered-mexican-food.

Lamden, Tim and Lucy Crossley, "FOUR planes have been involved in fatal near misses with drones at major British airports including Heathrow in the last month," *Daily Mail*, September 27, 2015. http://www.dailymail.co.uk/news/article-3251543/Drone-owners-forced-register-devices-tracking-database-four-near-misses-aircraft-past-month-alone.html.

Lovelace, Berkeley, Jr., "The future arrives? Amazon's Prime Air completes its first drone delivery," *CNBC*, December 14, 2016. http://www.cnbc.com/2016/12/14/the-future-arrives-amazons-prime-air-completes-its-first-drone-delivery.html.

Mac, Ryan, Frank Bi and Heng Shao, "World's Largest Drone Manufacturer DJI Seeking to Raise at $10 Billion Valuation," *Forbes*, April 14, 2015. http://www.forbes.com/sites/ryanmac/2015/04/14/worlds-largest-drone-manufacturer-dji-seeking-to-raise-at-10-billion-valuation/.

Mackenzie, Craig and Mark Duell, "'We hacked U.S. drone': Iran claims it electronically hijacked spy aircraft's GPS and tricked aircraft into landing on its soil," *Daily Mail*, December 19, 2001. http://www.dailymail.co.uk/news/article-2075157/Iran-claims-hacked-US-spy-planes-GPS-guided-aircraft-ground.html.

Madhani, Aamer, "Cleric al-Awlaki dubbed 'bin Laden of the Internet," *USA Today*, September 30, 2011. http://usatoday30.usatoday.com/news/nation/2010-08-25-1A_Awlaki25_CV_N.htm.

"Man, 26, charged in model airplane plot to bomb Pentagon," *DIY Drones*, September 28, 2011. http://diydrones.com/forum/topics/man-26-charged-in-model-airplane-plot-to-bomb-pentagon.

Markoff, John, "How Many Computers to Identify a Cat? 16,000," *New York Times*, June 25, 2012. http://www.nytimes.com/2012/06/26/technology/in-a-big-network-of-computers-evidence-of-machine-learning.html.

——, "Researchers Announce Advance in Image-Recognition Software," *New York Times*, November 14, 2014. http://www.nytimes.com/2014/11/18/science/researchers-announce-breakthrough-in-content-recognition-software.html.

Masi, Alessandria, "Hezbollah Allegedly Using Drones Against Al Qaeda In Battle For Qalamoun," *International Business Times*, May 12, 2015.

Mazzetti, Mark, Robert F. Worth and Eric Lipton, "Bomb Plot Shows Key Role Played By Intelligence," *New York Times*, October 31, 2010. http://www.nytimes.com/2010/11/01/world/01terror.html.

McLeary, Paul, "Marines extend Afghan deployment of cargo UAV," *Marine Times*, May 9, 2012. http://www.marinecorpstimes.com/news/2012/05/defense-marines-extend-kmax-afghan-deployment-050912/.

Mezzofiore, Gianluca, "Al-Qaeda in Yemen ideological leader Ibrahim al-Rubaish killed in drone strike," *International Business Times*, April 14, 2015. http://www.ibtimes.co.uk/al-qaeda-yemen-ideological-leader-ibrahim-al-rubaish-killed-drone-strike-1496395.

Michelzon, Tomer, "New Video Shows ISIS Using Drones to Plan Battles," *Vocativ*, August 25, 2014. http://www.vocativ.com/world/syria-world/new-video-shows-isis-using-drones-plan-battles/.

"Middle East Crisis: Facts and Figures," *BBC News*, August 31, 2006. http://news.bbc.co.uk/2/hi/middle_east/5257128.stm.

"The Military's New Weapon: Mini Spy Robots You Throw Like Grenades," *The Week*, March 23, 2012. http://theweek.com/article/index/226011/the-militarys-new-weapon-mini-spy-robots-you-throw-like-grenades.

Miller, Claire Cain, "With a Push from Google, California Legalizes Driverless Cars," *New York Times*, September 25, 2012. http://bits.blogs.nytimes.com/2012/09/25/with-a-push-from-google-california-legalizes-driverless-cars/.

Miller, Greg, "Plan for hunting terrorists signals U.S. intends to keep adding names to kill lists," *Washington Post*, October 23, 2012. https://www.washingtonpost.com/world/national-security/plan-for-hunting-terrorists-signals-us-intends-to-keep-adding-names-to-kill-lists/2012/10/23/4789-b2ae-18b3-11e2-a55c-39408fbe6a4b_story.html.

Miller, Greg and Bob Woodward, "Secret memos reveal explicit nature of U.S., Pakistan agreement on drones," *Washington Post*, October 24, 2013. https://www.washingtonpost.com/world/national-security/top-pakistani-leaders-secretly-backed-cia-drone-campaign-secret-documents-show/2013/10/23/15e6b0d8-3beb-11e3-b6a9-da62c264f40e_story.html.

"Mini army drones developed," *Reuters*, March 10, 2015. http://www.reuters.com/article/2015/03/10/us-poland-mini-drones-idUSKBN0M610P20150310.

Mothana, Ibrahim, "How Drones Help Al Qaeda," *New York Times*, June 13, 2012. http://www.nytimes.com/2012/06/14/opinion/how-drones-help-al-qaeda.html.

Mullen, Jethro, "Al Qaeda's second in command killed in Yemen strike; successor named," *CNN*, June 16, 2015. http://www.cnn.com/2015/06/16/middleeast/yemen-aqap-leader-killed/.

"Mumbai Attacks: Terrorists Monitored British Websites Using BlackBerry Phones," *Telegraph*, December 1, 2008, http://www.telegraph.co.uk/news/worldnews/asia/india/3534599/Mumbai-attacks-Terrorists-monitored-coverage-on-UK-websites-using-BlackBerry-phones-bombay-india.html.

Murray, Rebecca, "Anger at US drone war continues in Yemen," *Al Jazeera*, June 7, 2013. http://www.aljazeera.com/indepth/features/2013/06/201365122319329623.html.

Musharbash, Yassin, "A New Path for Al-Qaida: Zawahiri Confirmed as Bin Laden's Successor," *Der Spiegel*, June 16, 2001. http://www.spiegel.de/international/world/a-new-path-for-al-qaida-zawahiri-confirmed-as-bin-laden-s-successor-a-768849.html.

Nakashima, Ellen, "Officials: surveillance programs foiled more than 50 terrorist plots," *Washington Post*, June 18, 2013. https://www.washington-post.com/world/national-security/officials-surveillance-programs-foiled-more-than-50-terrorist-plots/2013/06/18/d657cb56-d83e-11e2-9df4-895344c13c30_story.html.

Nakashima, Ellen and Craig Whitlock, "With Air Force's Gorgon Drone 'we can see everything,'" *Washington Post*, January 2, 2011. http://www.washingtonpost.com/wp-dyn/content/article/2011/01/01/AR2011010102690.html.

"Nasrallah Hits Out at Government," *Al Jazeera*, May 8, 2008, http://english.aljazeera.net/NR/exeres/EB1FBB50-7FF6-4F98-B646-B2C795657F02.htm.

"Nasrallah Wins the War," *The Economist*, August 17, 2006, http://www.economist.com/opinion/displaystory.cfm?story_id=7796790.

Nelson, Bryn, "Checkers computer becomes invincible," *NBC News*, July 19, 2007. http://www.nbcnews.com/id/19839044/ns/technology_and_science-innovation/t/checkers-computer-becomes-invincible/#.Uc86KvnVCSo.

"No Military Solution to Iraq, U.S. General Says," *CNN*, March 9, 2007, http://www.cnn.com/2007/WORLD/meast/03/08/iraq.petraeus/index.html.

Pagliery, Jose, "Inside the $2 billion ISIS war machine," *CNN*, December 11, 2015. http://money.cnn.com/2015/12/06/news/isis-funding/.

"Parcel bomb plotters 'used dry run,' say US officials," *BBC News*, November 2, 2010. http://www.bbc.co.uk/news/world-us-canada-11671377.

"Paris drones: New wave of alerts," *BBC News*, March 4, 2015. http://www.bbc.com/news/world-europe-31725302.

Patel, Romil, "War on Isis: 305 jihadists killed by RAF drones in Iraq with 'no known civilian casualties,'" *International Business Times*, November 26, 2015. http://www.ibtimes.co.uk/war-isis-305-jihadists-killed-by-raf-drones-iraq-no-known-civilian-casualties-1530517.

Perry, Tom and Dan Williams, "Israeli drone strike in Syria kills two near frontier: Hezbollah's al-Manar TV," *Reuters*, July 29, 2015. http://www.reuters.com/article/us-mideast-crisis-syria-attack-idUSKCN0Q311T20150729.

Pfeffer Anshel, and Yanir Yagna, "Iron Dome successfully intercepts Gaza rocket for first time," *Haaretz*, April 7, 2011. http://www.haaretz.com/

news/diplomacy-defense/iron-dome-successfully-intercepts-gaza-rocket-for-first-time-1.354696.

Pfeiffer, Eric, "DARPA Unveils Robotic Mule," *Yahoo! News*, September 10, 2012. http://news.yahoo.com/darpa-unveils-robotic-mule.html.

Pilkington, Ed and Ewan MacAskill, "Obama's drone war a 'recruitment tool' for Isis, say US air force whistleblowers," *Guardian*, November 18, 2015. http://www.theguardian.com/world/2015/nov/18/obama-drone-war-isis-recruitment-tool-air-force-whistleblowers.

Pincus, Walter, "Air Force to Train More Remote than Actual Pilots," *Washington Post*, August 11, 2009. http://www.washingtonpost.com/wp-dyn/content/article/2009/08/10/AR2009081002712.html.

Pitzke, Marc, "How Drone Pilots Wage War," *Der Spiegel*, March 12, 2010. http://www.spiegel.de/international/world/0,1518,682420,00.html.

Popper, Ben, "Drone maker DJI nabs $75 million in funding at a $10 billion valuation," *The Verge*, May 6, 2015. http://www.theverge.com/2015/5/6/8554429/dji-75-million-funding-investment-accel-10-billion-valuation.

"Putin meets angry Beslan mothers," *BBC News*, September 2, 2005. http://news.bbc.co.uk/2/hi/europe/4207112.stm.

Quinn, Ben, "UK police see spike in drone incidents," *Guardian*, October 11, 2015. http://www.theguardian.com/technology/2015/oct/11/drone-incidents-reported-to-uk-police-on-the-rise.

Rawnsley, Adam, "Darpa's Cheetah-Bot Designed to Chase Human Prey," *Wired*, February 25, 2011. http://www.wired.com/dangerroom/2011/02/darpas-cheetah-bot-designed-to-chase-human-prey/.

——, "It's a Drone's World. We Just Live in it," *Wired*, November 28, 2011. http://www.wired.com/dangerroom/2011/11/drone-world/?pid=858.

Reidel, Bruce, "Al-Qaida's Hadramawt emirate," *Brookings*, July 12, 2015. http://www.brookings.edu/blogs/markaz/posts/2015/07/12-al-qaeda-yemen-emirate-saudi-riedel.

"Researchers use spoofing to 'hack' a drone." *BBC News*, June 29, 2012. http://www.bbc.com/news/technology-18643134.

Risen, James and David Johnston, "Threats and Responses: Hunt for Al Qaeda; Bush Has Widen Authority of C.I.A. to Kill Terrorists," *New York Times*, December 15, 2002. http://www.nytimes.com/2002/12/15/world/threats-responses-hunt-for-al-qaeda-bush-has-widened-authority-cia-kill.html?pagewanted=all.

Ritter, Karl, and Lorne Cook, "Death toll rises to 130 following Paris attacks," *Toronto Star*, November 20, 2015. http://www.thestar.com/news/world/2015/11/20/third-body-found-at-scene-of-paris-attacks-police-raid.html.

"Robocopter arrives," *The Economist*, September 15–21, 2012. http://www.economist.com/node/21562897.

"Robodiptera," *The Economist*, May 4, 2013, p. 77.

Sasso, Brendan, "Hollywood wants drones for filmmaking," *The Hill*, January 25, 2013.

"Saudi Arabia Says It Foiled Attack on U.S. Embassy, Arrested ISIS Supporters," *NBC News*, April 28, 2015. http://www.nbcnews.com/news/world/saudi-arabia-says-it-foiled-attack-u-s-embassy-arrested-n349606.

Savage, Charlie, "Secret U.S. Memo Made Legal Case to Kill a Citizen," *New York Times*, October 8, 2011. http://www.nytimes.com/2011/10/09/world/middleeast/secret-us-memo-made-legal-case-to-kill-a-citizen.html?hp=&pagewanted=all.

Schmidt, Michael S. and Eric Schmitt, "Pentagon Confronts a New Threat from ISIS: Exploding Drones," *New York Times*, October 11, 2016. https://www.nytimes.com/2016/10/12/world/middleeast/iraq-drones-isis.html?hp&action=click&pgtype=Homepage&clickSource=story-heading&module=first-column-region&region=top-news&WT.nav=top-news&_r=1.

Schneider, Howard, "Hezbollah rearms away from border," *Washington Post*, January 23, 2010. http://articles.washingtonpost.com/2010-01-23/world/36923442_1_hezbollah-litani-river-judith-palmer-harik.

"Scores Killed in Mumbai Attacks," *Al Jazeera*, November 27, 2008, http://english.aljazeera.net/news/asia/2008/11/2008112617472965818.html.

"Secret US drone base in Saudi Arabia revealed," *BBC News*, February 6, 2013. http://www.bbc.com/news/world-middle-east-21361992.

Shachtman, Noah, "Flying Spy Surge: Surveillance Missions Over Afghanistan Quadruple," *Wired*, October 19, 2011. http://www.wired.com/dangerroom/2011/10/flying-spy-surge/.

Shane, Scott, "Drone Strikes Reveal Uncomfortable Truth: U.S. Is Often Unsure About Who Will Die," *New York Times*, April 23, 2015. http://www.nytimes.com/2015/04/24/world/asia/drone-strikes-reveal-uncomfortable-truth-us-is-often-unsure-about-who-will-die.html?_r=1.

Shane, Scott, and Eric Schmitt, "One Drone Victim's Trail from Raleigh to Pakistan," *New York Times*, May 22, 2013. http://www.nytimes.com/2013/05/23/us/one-drone-victims-trail-from-raleigh-to-pakistan.html.

Shear, Michael D. and Michael S. Schmidt, "White House Drone Crash Described as a U.S. Worker's Drunken Lark," *New York Times*, January 27, 2015. http://www.nytimes.com/2015/01/28/us/white-house-drone.html.

Singer, Peter W., "Do Drones Undermine Democracy?," *New York Times*, January 21, 2012. http://www.nytimes.com/2012/01/22/opinion/sunday/do-drones-undermine-democracy.html.

Sly, Liz, "Al-Qaeda disavows any ties with radical Islamist ISIS group in Syria, Iraq," *Washington Post*, February 3, 2014. https://www.washingtonpost.com/world/middle_east/al-qaeda-disavows-any-ties-with-radical-islamist-isis-group-in-syria-iraq/2014/02/03/2c9afc3a-8cef-11e3-98ab-fe5228217bd1_story.html.

Sofer, Roni, "Israel in favor of extending Gaza lull," *Ynet News*, December 13, 2008. http://www.ynetnews.com/articles/0,7340,L-3637877,00.html.

——, "Israel in favor of extending Gaza lull," *Ynet News*, November 5, 2011. http://www.maannews.net/eng/ViewDetails.aspx?ID=533909.

Specia, Megan, "Who are the Americans who have been killed by U.S. drone strikes?," *Mashable*, April 25, 2015. http://mashable.com/2015/04/25/americans-killed-us-drone-strikes/#Vdi3cEJ35GqT.

Starr, Barbara and Catherine E. Shoichet, "Russian plane crash: U.S. intel suggests ISIS bomb brought down jet," *CNN*, November 4, 2015.

Stuster, J. Dana, "Why Hezbollah's New Missiles Are a Problem for Israel," *Foreign Policy*, January 3, 2014. http://foreignpolicy.com/2014/01/03/why-hezbollahs-new-missiles-are-a-problem-for-israel/.

Teinowitz, Ira, "Hollywood to the FAA: Let Us Use Drones!," *The Wrap*, February 5, 2013. http://www.thewrap.com/movies/column-post/hollywood-faa-let-us-use-drones-76011.

Terdiman, Daniel, "Drone dogfights by 2015? U.S. Navy preps futuristic combat," *CNET*, June 21, 2012. http://news.cnet.com/8301-13576_3-57457501-315/drone-dogfights-by-2015-u.s-navy-preps-for-futuristic-combat/.

Thompson, Mark, "Iron Dome: A Missile Shield That Works," *Time*, November 19, 2012. http://nation.time.com/2012/11/19/iron-dome-a-missile-shield-that-works/.

——, "U.S. Bombing of ISIS Oil Facilities Showing Progress," *Time*, December 13, 2015. http://time.com/4145903/islamic-state-oil-syria.

"Tiny airplanes and subs from University of Florida laboratory could be next hurricane hunters," *University of Florida News*, June 4, 2013. http://news.ufl.edu/2013/06/04/hurricane-drones/.

"U.S. aircraft strike ISIS drone in Iraq: officials," *Al Arabiya*, March 18, 2015. http://english.alarabiya.net/en/News/middle-east/2015/03/18/U-S-aircraft-strike-ISIS-drone-in-Iraq-officials-.html.

"U.S. Navy SEALs fail to capture al-Shabaab commander," *Al Arabiya*, October 7, 2013. http://english.alarabiya.net/en/News/africa/2013/10/07/U-S-raid-in-Somalia-targeted-al-Shabaab-commander-.html.

Van Dyk, Jere, "Who were the 4 U.S. citizens killed in drone strikes?," *CBS News*, May 23, 2013. http://www.cbsnews.com/news/who-were-the-4-us-citizens-killed-in-drone-strikes/.

Vardi, Nathan, "Is al Qaeda Bankrupt?," *Forbes*, February 11, 2010. http://www.forbes.com/forbes/2010/0301/terrorism-funds-finance-osama-al-qaeda-bankrupt.html.

Vick, Karl, "Spy Fail: Why Iran Is Losing Its Covert War with Israel," *Time*, February 13, 2013. http://world.time.com/2013/02/13/spy-fail-why-iran-is-losing-its-covert-war-with-israel/.

"Video shows capture of Abu Anas al-Libi," *Al Arabiya*, February 11, 2014. http://english.alarabiya.net/en/News/middle-east/2014/02/11/Video-shows-capture-of-most-wanted-terrorist-Anas-al-Libi.html.

Wagstaff, Keith, "FAA Misses Deadline for Creating Drone Regulations," *NBC News*, October 1, 2015. http://www.nbcnews.com/tech/innovation/faa-misses-deadline-creating-drone-regulations-n437016.

Warrick, Joby, "Use of weaponized drones by ISIS spurs terrorism fears," *Washington Post*, February 21, 2016. https://www.washingtonpost.com/world/national-security/use-of-weaponized-drones-by-isis-spurs-terrorism-fears/2017/02/21/9d83d51e-f382–11e6–8d72–263470bf0401_story.html? utm_term=.5ebb8e18d16a.

"Washington Post-ABC News Poll," *Washington Post*, February 4, 2012. http://www.washingtonpost.com/wp-srv/politics/polls/postabcpoll_020412. html.

Watt, Nicholas, "The 'kill list': RAF drones have been hunting UK jihadis for months," *Guardian*, September 8, 2015. http://www.theguardian.com/ uk-news/2015/sep/08/drones-uk-isis-members-jihadists-syria-kill-list-ministers.

Weiser, Benjamin and Michael S. Schmidt, "Qaeda Suspect Facing Trial in New York Over Africa Embassy Bombings Dies," *New York Times*, January 3, 2015. http://www.nytimes.com/2015/01/04/us/politics/qaeda-suspect-facing-trial-in-new-york-dies-in-custody.html?_r=0.

Weiss, Caleb, "AQAP shows fighting in strategic Yemeni city in new video," *Long War Journal*, November 11, 2015. http://www.longwarjournal.org/ archives/2015/11/aqap-shows-fighting-in-strategic-yemeni-city-in-new-video.php.

Whitlock, Craig, "Italy convicts 22 CIA operatives, U.S. Air Force colonel in rendition case," *Washington Post*, November 5, 2009. http://www.washington post.com/wp-dyn/content/article/2009/11/04/AR2009110404525.html.

——, "More Air Force drones are crashing than ever as mysterious new problems emerge," *Washington Post*, January 20, 2016. https://www. washingtonpost.com/news/checkpoint/wp/2016/01/19/more-u-s-military-drones-are-crashing-than-ever-as-new-problems-emerge/.

"Why Soldiers Hate the Raven UAV," *Military.com*, May 29, 2012. http:// www.military.com/video/aircraft/pilotless-aircraft/why-soldiers-hate-the-raven-uav/1661802396001/.

Wilson, Chris, "Jeopardy, Schmeopardy," *Slate*, February 15, 2011. http:// www.slate.com/articles/health_and_science/science/2011/02/jeopardy_ schmeopardy.html.

Wingfield, Nick, "New F.A.A. Report Tallies Drone Sightings, Highlighting Safety Issues," *New York Times*, November 26, 2014. http://bits.blogs.nytimes.com/ 2014/11/26/new-f-a-a-report-tallies-drone-sightings-highlighting-safety-issues/.

Wingfield, Nick and Somini Sengupta, "Drones Set Sights on U.S. Skies," *New York Times*, February 17, 2012. http://www.nytimes.com/2012/02/18/ technology/drones-with-an-eye-on-the-public-cleared-to-fly.html? pagewanted=all.

Wright, Robin, "Strikes Are Called Part of Broad Strategy," *Washington Post*, July 16, 2006. http://www.washingtonpost.com/wp-dyn/content/article/ 2006/07/15/AR2006071500957_pf.html.

"Yemen-based al Qaeda group claims responsibility for parcel bomb plot," *CNN*, November 6, 2010. http://edition.cnn.com/2010/WORLD/meast/ 11/05/yemen.security.concern/?hpt=T2.

"Yemen crisis: US troops withdraw from air base," *BBC News*, March 22, 2015. http://www.bbc.com/news/world-middle-east-32000970.

Yeung, Peter, "UK air strikes kill 1,000 Isis fighters in Iraq and Syria but no civilians, officials claim," *Independent*, April 30, 2016. http://www. independent.co.uk/news/uk/home-news/iraq-syria-air-strikes-civilians-casualties-killed-isis-daesh-islamic-state-air-wars-a7008276.html.

Ying, Wang, "DJI sees jump in revenue," *China Daily*, December 13, 2016. http://www.chinadaily.com.cn/business/tech/2016–12/13/content_27649387.htm.

Young, Leslie, "Drones increasingly buzzing too close to Canadian airports: reports," *Global News*, March 10, 2015. http://globalnews.ca/news/1872447/drones-increasingly-buzzing-too-close-to-canadian-airports-reports/.

TRADE JOURNAL AND MAGAZINE ARTICLES

Ackerman, Evan, "ATLAS DRC Robot is 75 Percent New, Completely Unplugged," *IEEE Spectrum*, January 20, 2015. http://spectrum.ieee.org/automaton/robotics/military-robots/atlas-drc-robot-is-75-percent-new-completely-unplugged.

——, "Dutch Police Training Eagles to Take Down Drones," *IEEE Spectrum*, February 1, 2016. http://spectrum.ieee.org/automaton/robotics/drones/dutch-police-training-eagles-to-take-down-drones.

"A Laser Phalanx?," *Defense Industry Daily*, April 23, 2009. http://www.defensei ndustrydaily.com/a-laser-phalanx-03783/.

Anderson, Chris, "The DIY Drones Mission (aka The Five Rules)", *DIY Drones*, January 4, 2008. http://diydrones.com/profiles/blog/show?id=705844: BlogPost:17789.

——, "A newbie's guide to UAVs," *DIY Drones*, March 28, 2009. http:// diydrones.com/profiles/blogs/a-newbies-guide-to-uavs.

"Armed, Aware and Dangerous: the Top Five Military Robots," *Army-Technology.com*, February 27, 2012. http://www.army-technology.com/features/featurearmed-aware-and-dangerous-the-top-five-military-robots/.

Atherton, Kelsey D., "Watch a Hacker Take Over a Drone Remotely," *Popular Science*, August 18, 2015. http://www.popsci.com/watch-hacker-down-drone-button-press.

——, "Israeli Contractor Rafael Shows Off Anti-Drone Laser in Korea," *Popular Science*, October 20, 2015. http://www.popsci.com/israels-rafael-shows-off-anti-drone-laser-in-korea.

Atwood, Tom, and Jonathan Klein, "Vecna's Battlefield Extraction-Assist Robot BEAR," *Robot*, April 25, 2007. http://www.botmag.com/articles/04–25-07_vecna_bear.shtml.

Axe, David, "Buyer's Remorse: How Much Has the F-22 Really Cost?," *Wired*, December 14, 2011. http://www.wired.com/2011/12/f-22-real-cost/.

Binnie, Jeremy, "General Atomics confirms UAE Predator delivery," *IHS Jane's 360*, February 17, 2017. http://www.janes.com/article/67797/general-atomics-confirms-uae-predator-delivery.

Blanford, Nicholas, "Hizbullah airstrip revealed," *IHS Jane's 360*, April 23, 2015. http://www.janes.com/article/50922/hizbullah-airstrip-revealed.

Carey, Bill, "General Atomics Plans Predator XP Deliveries to UAE," *AIN Online*, November 9, 2015. http://www.ainonline.com/aviation-news/defense/2015-11-09/general-atomics-plans-predator-xp-deliveries-uae.

——, "General Atomics Predator C Avenger ER Makes First Flight," *AIN Online*, November 11, 2016. http://www.ainonline.com/aviation-news/defense/2016-11-11/general-atomics-predator-c-avenger-er-makes-first-flight.

Cenciotti, David, "Iran's New Spy Drone Is an Israeli Hermes 450/Watchkeeper Clone Capable of Carrying Missiles," *The Aviationist*, September 25, 2012. http://theaviationist.com/2012/09/25/shahed129/.

——, "Hamas Flying An Iranian-Made Armed Drone Over Gaza," *The Aviationist*, July 14, 2014. http://theaviationist.com/2014/07/14/ababil-over-israel/.

——, "ISIS Surveillance Drone Is Only an Amateurish Remote-Controlled Quad-Copter," *The Aviationist*, September 2, 2014.

"Civilian drones and the legal issues surrounding their use," *Wilde Beuger Solmecke*, February 18, 2014. https://www.wbs-law.de/internetrecht/civilian-drones-legal-issues-surrounding-use-50459/.

Clark, Colin, "Gorgon Stare Blinks A Lot; Testers Say Don't Field Til Fixed," *DoDBuzz.com*, January 24, 2011. http://www.dodbuzz.com/2011/01/24/gordon-stare-blinks-a-lot-testers-say-dont-field-til-fixed/.

Condliffe, Jamie, "A 100-Drone Swarm, Dropped from Jets, Plans Its Own Moves," *Technology Review*, January 10, 2017. https://www.technologyre-view.com/s/603337/a-100-drone-swarm-dropped-from-jets-plans-its-own-moves/.

Crane, David, "Anti-Sniper/Sniper Detection/Gunfire Detection Systems at a Glance," *Defense Review*, July 19, 2006. http://www.defensereview.com/anti-snipersniper-detectiongunfire-detection-systems-at-a-glance/.

Davey, Tucker, "Explainable AI: a discussion with Dan Levy," *Future of Life Institute*, September 27, 2017. https://futureoflife.org/2017/09/27/explain-able-ai-a-discussion-with-dan-weld/.

Diaz, Jesus, "Robot horse gets first taste of real-world action with the US Marines," *Sploid*, July 14, 2014. http://sploid.gizmodo.com/big-dog-robothas-been-deployed-by-the-us-army-for-the-1604647456.

Dogaru, Traian, and Calvin Le, "Validation of Xpatch Computer Models for Human Body Radar Signature," *Army Research Laboratory*, March 2008, http://www.arl.army.mil/arlreports/2008/ARL-TR-4403.pdf.

"Drone Laws By Country," *UAV Systems International*. https://uavsystemsin-ternational.com/drone-laws-by-country/?v=7516fd43adaa.

"Drone maker DJI rumored to be planning for an IPO in 2017." *AllChinaTech*, February 22, 2016. http://www.allchinatech.com/drone-maker-dji-rum ored-planning-ipo-2017/.

"Drones: Reporting for Work," *Goldman Sachs*. http://www.washingtonpost. com/sf/brand-connect/goldman-sachs/drones-reporting-for-work/.

Dsouza, Larkins, "RQ-170 Sentinel 'Beast of Kandahar,'" *Defense Aviation*, December 26, 2009. http://www.defenceaviation.com/2009/12/rq-170-sentinel-beast-of-kandahar-confirmed-by-us-airforce.html.

Dunnigan, James, "Switchblade Enters Service," *Strategy Page*, September 24, 2011. http://www.strategypage.com/dls/articles/Switchblade-Enters-Service-9-24-2011.asp.

Fingas, Jon, "Google lands patent for automatic object recognition in videos, leaves no stone untagged," *Engadget*, August 28, 2012. http://www.engadget.com/2012/08/28/google-lands-patent-for-automatic-object-recognition-in-videos/.

Finkelstein, Robert, "Military Robotics: Malignant Machines or the Path to Peace?," *Robotic Technology Inc.*, January 2010. http://www.robotic technologyinc.com/images/upload/file/Presentation%20Military%20 Robotics%20Overview%20Jan%2010.pdf.

Freedburg Jr., Sydney J., "Centaur Army: Bob Work, Robotics, & The Third Offset Strategy," *Breaking Defense*, November 9, 2015. http://breakingdefense.com/ 2015/11/centaur-army-bob-work-robotics-the-third-offset-strategy/?utm_ campaign=Breaking+Defense + Daily + Digest&utm_source=hs_email& utm_medium=email&utm_content=23591130&_hsenc=p2ANqtz-9ju6Qz 2QE7JdLwVrBALonZrml_S21s5iD5Fv-WoG7tqoduO4SJmKAC7Bp_s9qoh9 ppiCRpJz7NFQdggIKsEWFMCkIorg&_hsmi=23591130.

"French Drone Regulations – Updated as of 6/23/2015," *The Drone Info*, June 23, 2015. http://www.thedroneinfo.com/french-drone-regulations/.

Golson, Jordan, "Welcome to the World, Drone-Killing Laser Cannon," *Wired*, August 27, 2015. http://www.wired.com/2015/08/welcome-world-drone-killing-laser-cannon/.

Green, Ollie, "Dragonfly Robotic Insect UAV is Freaking Cool," *Mobile*, November 7, 2012. http://www.mobilemag.com/2012/11/07/dragonfly-robotic-insect-uav-is-freaking-cool/.

Guizzo, Erico and Evan Ackerman, "How South Korea's DRC-HUBO Robot Won the DARPA Robotics Challenge," *IEEE Spectrum*, June 9, 2015. http:// spectrum.ieee.org/automaton/robotics/humanoids/how-kaist-drc-hubo-won-darpa-robotics-challenge.

Hill, David J., "Toy-Size Helicopter Drones Now on Surveillance Duty in Afghanistan," *SingularityHUB*, February 11, 2013. http://singularityhub. com/2013/02/11/toy-size-helicopter-drones-now-on-surveillance-duty-in-afghanistan/.

Hoffman, Mike, "PBS Features DARPA's ARGUS-IS," *Defensetech*, January 29, 2013. http://defensetech.org/2013/01/29/pbs-features-darpas-argus-is/.

——, "British soldiers flying nano helicopters in Afghanistan," *Defensetech*, February 5, 2013. http://defensetech.org/2013/02/05/british-soldiers-flying-nano-helicopters-in-afghanistan/.

Humphries, Matthew, "WASP: The Linux-powered flying spy drone that cracks Wi-Fi & GSM networks," *Geek.com*, July 29, 2011. http://www.geek.com/

articles/geek-pick/wasp-the-linux-powered-flying-spy-drone-that-cracks-wi-fi-gsm-netwokrs-20110729/.

Kelly, Kevin, "The future of AI? Helping human beings think smarter." *Wired*, December 3, 2014. http://www.wired.co.uk/magazine/archive/2014/12/features/brain-power/page/2.

Labella, Thomas H., Marco Dorigo, and Jean-Louis Deneubourg, "Division of Labor in a Group of Robots Inspired by Ants' Foraging Behavior," *ACM Transactions on Autonomous and Adaptive Systems 1 (1)*, pp. 4–25, 2006. http://www.swarm-bots.org/dllink.php?id=751&type=documents.

Lum, Zachary, "The Measure of MASINT," *Journal of Electronic Defense*, August 1, 1998. http://www.globalsecurity.org/intell/library/news/1998/08/MASINT.htm.

Matyszczyk, Chris, "Need to take down a drone? A munitions company offers firepower," *CNET*, August 23, 2015. http://www.cnet.com/news/company-markets-anti-drone-munitions/.

McLeary, Paul, "K-MAX Chugging Along in Afghanistan," *Aviation Week*, February 3, 2012. http://www.aviationweek.com/Blogs.aspx?plckBlogId=Blog:27ec4a53-dcc8–42d0-bd3a-01329aef79a7&plckController=Blog&plckBlogPage=BlogViewPost&newspaperUserId=27ec4a53-dcc8–42d0-bd3a-01329aef79a7&plckPostId=Blog%253a27ec4a53-dcc8–42d0-bd3a-01329aef79a7Post%253a32270b95-e2c6–4d57–8ebd-a9ed007f342c&plckScript=blogScript&plckElementId=blogDest.

Moubarak, Paul and Pinhas Ben-Tzvi, "Modular and reconfigurable mobile robotics," *Robotics and Autonomous Systems* vol. 60 (2012), 1648–1663.

Oswald, Ed, "This Anti-UAV Octocopter Uses a Ballistic Net Cannon to Disable Smaller Drones," *Digital Trends*, January 12, 2016. http://www.digitaltrends.com/cool-tech/bad-drones-beware-drone-coming-get/.

Pardesi, Manjeet Singh, "Unmanned Aerial Vehicles/Unmanned Aerial Combat Vehicles: Likely Missions and Challenges for the Policy-Relevant Future," *Air & Space Power Journal*, Fall 2005. http://www.airpower.au.af.mil/airchronicles/apj/apj05/fal05/pardesi.html.

Patil, Madhav, Tamer Abukhalil and Tarek Sobh, "Hardware Architecture Review of Swarm Robotics System: Self-Reconfigurability, Self-Reassembly, Self-Replication," *ISRN Robotics*, Volume 2013 (2013).

"Rise of the Machines," *Army-Technology.com*, May 21, 2008. http://www.army-technology.com/features/feature1951/.

Ruppert, Barb, "The Battlefield-Extraction-Assist Robot to Rescue Wounded on Battlefield," *MilitaryInfo.com*, November 22, 2010. http://www.militaryinfo.com/news_story.cfm?textnewsid=6556.

Sahin, Erol and Nigel R. Franks, "Measurement of Space: From Ants to Robots," *Proceedings of WGW 2002: EPSCRBBSRC International Workshop Biologically-Inspired Robotics*, pp. 241–247, Bristol, UK, August 14–16, 2002. http://www.swarm-bots.org/dllink.php?id=161&type=documents.

Sanborn, James K., "Beacon improves UAV cargo-delivery accuracy," *Marine Times*, July 8, 2012. http://www.marinecorpstimes.com/news/2012/07/marine-kmax-beacon-improves-uav-cargo-delivery-accuracy-070812w/.

Simonite, Tom, "A Brain-Inspired Chip Takes to the Sky," *MIT Technology Review*, November 4, 2014. http://www.technologyreview.com/news/532176/a-brain-inspired-chip-takes-to-the-sky/.

——, "Facebook Creates Software That Matches Faces Almost as Well as You Do," March 17, 2014. *MIT Technology Review*, http://www.technologyreview.com/news/525586/facebook-creates-software-that-matches-faces-almost-as-well-as-you-do/.

Smalley, David, "LOCUST: Autonomous, swarming UAVs fly into the future," *Office of Naval Research*, April 14, 2014. http://www.onr.navy.mil/Media-Center/Press-Releases/2015/LOCUST-low-cost-UAV-swarm-ONR.aspx.

Smith, Lisa, "Did ETFs Cause the Flash Crash?," *Investopedia. http://www.investopedia.com/articles/exchangetradedfunds/11/ets-flash-crash.asp.*

Smith, Mat, "Marines send its 'AlphaDog' robot to the farm," *Engadget*, December 29, 2015. http://www.engadget.com/2015/12/29/marines-send-its-alphadog-robot-to-the-farm/.

Sofge, Erik, "5 Robots We Should Deploy Right Now," *Popular Mechanics*, April 13, 2010. http://www.popularmechanics.com/technology/engineering/robots/robots-to-deploy-now.

——, "The DARPA Robotics Challenge Was a Bust," *Popular Science*, July 6, 2015. http://www.popsci.com/darpa-robotics-challenge-was-bust-why-darpa-needs-try-again.

Stirling, Timothy, James Robert, Jean-Christophe Zufferey and Dario Floreano, "Indoor Navigation with a Swarm of Flying Robots," *Proceedings of the 2012 IEEE International Conference on Robotics and Automation*, 2012. http://www.swarmanoid.org/upload/pdf/StirlingEtAl2012.pdf.

"Swarmanoid: Towards Humanoid Robotic Swarms," *Swarmanoid.org*. http://www.swarmanoid.org/index.php.

"Swarm-bots: Swarms of self-assembling artifacts," *Swarm-bots.org*. http://www.swarm-bots.org/.

Szondy, David, "Neither rain, nor fog, nor wind stops Boeing's laser weapon destroying targets," *Gizmag*, September 8, 2014. http://www.gizmag.com/boeing-laser-directed-energy-weapon-fog/33672/.

"Teal Group Predicts Worldwide UAV Market Will Total $91 Billion in Its 2014 UAV Market Profile and Forecast," *Teal Group Corporation*, July 17, 2014. http://www.tealgroup.com/index.php/about-teal-group-corporation/press-releases/118-2014-uav-press-release.

"The Drones Report: Market forecasts, regulatory barriers, top vendors, and leading commercial applications," *Business Insider*, May 27, 2015. http://www.businessinsider.com/uav-or-commercial-drone-market-forecast-2015-2.

Trianni, Vito and Marco Dorigo, "Emergent Collective Decisions in a Swarm of Robots," *2005 IEEE Swarm Intelligence Symposium (SIS 2005)*, pp. 241–248,

June 8–10, 2005. http://www.swarm-bots.org/dllink.php?id=692& type=documents.

Trimble, Stephen, "Lockheed Martin to Build the Mother of All Airborne Radars," *The DEW Line*, April 27, 2009. http://www. flightglobal.com/blogs/the-dewline/2009/04/lockheed-martin-to-build-the-m.html.

Ungerleider, Neal, "The Science Behind the Drone Terrorism Attack," *Fast Company*, September 29, 2011. http://www.fastcompany.com/1783721/ science-behind-drone-terrorism-attack.

"Unmanned Aircraft Systems Flight Plan," *United States Air Force*, May 18, 2009. http://www.scribd.com/doc/17312080/United-States-Air-Force-Unmanned-Aircraft-Systems-Flight-Plan-20092047-Unclassified.

"Unmanned K-MAX Wins Top Honors, USMC Praise," *HeliHub*, January 9, 2013. http://helihub.com/2013/01/09/unmanned-k-max-wins-top-innovation-honors-usmc-praise/.

"U.S. Army Awards AeroVironment $5.1 Million Order for Switchblade Loitering Munitions System and Services," *AeroVironment*, May 23, 2012. http://www.avinc.com/resources/press_release/u.s._army_awards _aerovironment_5.1_million_order_for_switchblade_loitering_.

"US Military Bringing a Switchblade to A Gun Fight," *Defense Industry Daily*, September 13, 2012. http://www.defenseindustrydaily.com/us-army-brings-a-switchblade-to-a-gun-fight-07071/.

Warwick, Graham, "Airship Programs – Not So Buoyant, Says GAO," *Ares*, October 27, 2012. http://aviationweek.com/blog/airship-programs-not-so-buoyant-says-gao.

Watson, Ben, "The Drones of ISIS," *Defense One*, January 12, 2017. http://www. defenseone.com/technology/2017/01/drones-isis/134542/.

Weiss, Caleb, "Islamic State uses drones to coordinate fighting in Baiji," *Long War Journal*, April 17, 2015. http://www.longwarjournal.org/ archives/2015/04/islamic-state-uses-drones-to-coordinate-fighting-in-baiji.php.

Yirka, Bob, "Makers of infamous BigDog robot unveil human version – PETMAN," *Phys.org*, November 1, 2011. http://phys.org/news/2011–11-makers-infamous-bigdog-robot-unveil.html.

TECHNICAL SPECIFICATIONS, MANUFACTURER INFORMATION, AND DEMONSTRATION VIDEOS

"Ababil (Swallow) Unmanned Aerial Vehicle," *Globalsecurity.org*. http://www. globalsecurity.org/military/world/iran/ababil.htm.

"About Us," *The Rabbit-Hole*, https://rabbit-hole.org/about/.

Alex, Dan, "IAI Heron/Machatz-1 Unmanned Aerial Vehicle," *Military Factory*, October 26, 2015. http://www.militaryfactory.com/aircraft/detail.asp? aircraft_id=823.

"ArduCopter User Group," *DIY Drones.* http://diydrones.com/group/arducopterusergroup.

"ArduPilot," *code.google.com.* https://code.google.com/p/ardupilot/.

"Army Orders 1,100 Recon Scout XT Robots from ReconRobotics," *Business Wire,* February 15, 2012. http://www.businesswire.com/news/home/20120215005395/en/Army-Orders-1100-Recon-Scout-XT-Robots.

"ARPAnet," *PCMAG.com.* http://www.pcmag.com/encyclopedia/term/37989/arpanet.

"AutoCAD Map 3D," *Autodesk.* http://www.autodesk.com/products/autodesk-autocad-map-3d/overview.

"Autonomous Real-Time Ground Ubiquitous Surveillance-Imaging System (ARGUS-IS)," *DARPA,* http://www.darpa.mil/Our_Work/I2O/Programs/Autonomous_Real-time_Ground_Ubiquitous_Surveillance-Imaging_System_(ARGUS-IS).aspx.

"BigDog Overview (video)," *Boston Dynamics,* http://www.youtube.com/watch?v=cNZPRsrwumQ.

"BigDog – The Most Advanced Rough-Terrain Robot on Earth," *Boston Dynamics,* http://www.bostondynamics.com/robot_bigdog.html.

"Burrito Bomber," *Darwin Aerospace.* http://www.darwinaerospace.com/burritobomber.

"Center for Neural and Emergent Systems," *HRL.* http://www.hrl.com/laboratories/cnes/cnes_main.html.

"Cheetah Robot runs 28.3 mph; a bit faster than Usain Bolt," *Boston Dynamics,* September 5, 2012. http://www.youtube.com/watch?v=chPanW0QWhA&list=UU7vVhkEfw4nOGp8TyDk7RcQ&index=3.

"Counter Rocket, Artillery, and Mortar (C-RAM)," *globalsecurity.org.* http://www.globalsecurity.org/military/systems/ground/cram.htm.

"Decibel Levels of Everyday Sounds," *informationxchange.* http://www.stuartxchange.com/Decibels.html.

"Download the Arduino Software," *Arduino.cc.* http://arduino.cc/en/Main/Software.

"Dynamic Robot Manipulation," *Boston Dynamics,* March 1, 2013. http://www.youtube.com/user/BostonDynamics?feature=watch.

"E-Flite F-86 Sabre 15 Ducted Fan Jet ARF," *A Main Hobbies.* http://www.amainhobbies.com/product_info.php/cPath/3_516_2077_693/products_id/173356/n/E-Flite-F-86-Sabre-15-Ducted-Fan-Jet-ARF?utm_source=GoogleBase&utm_medium=cpc&utm_campaign=Product-Feeds&source=google_ext&gclid=CPirx5Chp7cCFYyZ4AodLS4AYA.

"F-22 Raptor," *U.S. Air Force,* September 23, 2015. http://www.af.mil/AboutUs/FactSheets/Display/tabid/224/Article/104506/f-22-raptor.aspx.

"F/A-18 *Hornet* Strike Fighter," *United States Navy Fact File,* May 26, 2009. http://www.navy.mil/navydata/fact_display.asp?cid=1100&tid=1200&ct=1.

"Fido XT Explosives Trace Detector," *FLIR Systems.* http://www.flir.com/threatdetection/display/?id=63353.

"Flying-Cam and Bond 007 'Skyfall,'" *Flying-Cam*, April 24, 2012. http://flying-cam.com/en/news.php.

"Gorgon Stare," *Globalsecurity.org*. http://www.globalsecurity.org/intell/systems/gorgon-stare.htm.

Greenberg, Andy, "Hacker says he can hijack a $35k police drone a mile away," *Wired*, March 2, 2016.https://www.wired.com/2016/03/hacker-says-can-hijack-35k-police-drone-mile-away/.

"Ground Robots – 510 PackBot," *iRobot*. http://www.irobot.com/gi/ground/510_PackBot

"Hezbollah Displays Iranian Fajr-5 Missile," posted on *YouTube* April 2, 2013: http://www.youtube.com/watch?v=CoacPETi26k and April 6, 2013: http://www.youtube.com/watch?v=c9Ad6NYxQ60.

"High Performance Hydraulics for Industrial Applications," *Vecna*, http://www.vecna.com/robotics/multimedia/downloads/high_performance_hydraulics_for_industrial_applications.pdf.

"Integrated Sensor Is Structure (ISIS)," *DARPA Strategic Technology Office*, http://www.darpa.mil/Our_Work/STO/Programs/Integrated_Sensor_is_Structure_%28ISIS%29.aspx.

"iRobot PackBot 510 with Engineer Kit," *iRobot*, http://www.ulkem.com.tr/html/urunler/robotlar/PackBot510engineer/ppdf.pdf.

"Iron Dome," *Rafael Advanced Defense Systems*. http://www.rafael.co.il/Marketing/186–1530-en/Marketing.aspx.

Kamkar, Samy, "SkyJack." http://samy.pl/skyjack/.

"List All Manufacturers," *UAVGlobal*. http://www.uavglobal.com/list-of-manufacturers/.

"Lockheed Martin RQ-170 Sentinel Unmanned Aerial Vehicle," *Miltaryfactory.com*, December 12, 2011. http://www.militaryfactory.com/aircraft/detail.asp?aircraft_id=896.

"MK 15 – Phalanx Close-In Weapons System (CIWS)," *United States Navy Fact File*. http://www.navy.mil/navydata/fact_display.asp?cid=2100&tid=487&ct=2.

"Mohajer (UAV)," *Globalsecurity.org*. http://www.globalsecurity.org/military/world/iran/mohajer.htm.

"Move Like You Think, A Thought Made Invisible," *Flying-Cam*. http://www.flying-cam.com/#.

"MQ-1 Predator," *Deagel.com*, October 28, 2015. http://www.deagel.com/Unmanned-Combat-Air-Vehicles/MQ-1-Predator_a000517002.aspx.

"MQ-1B Predator," *U.S. Air Force Fact Sheet*, September, 2015. http://www.af.mil/AboutUs/FactSheets/Display/tabid/224/Article/104469/mq-1b-predator.aspx.

"MQ-9 Reaper," *U.S. Air Force Fact Sheet*, September 23, 2015. http://www.af.mil/AboutUs/FactSheets/Display/tabid/224/Article/104470/mq-9-reaper.aspx.

"NBS MANTIS Air Defense Protection System, Germany," *Army-technology.com*. http://www.army-technology.com/projects/mantis/.

"Official FAA Approval for Flying-cam 3.0 SARAH," *Flying-Cam*, October 15, 2014. http://flying-cam.com/en/news.php?id=133&PHPSESSID=efe85 f3c35a7ad5c76fdda3a35ab92e8.

"PackBot Tactical Robot," *Defense Update*, http://defense-update.com/products/p/pacbot.htm.

"Packet Switching," *PCMAG.com*. http://www.pcmag.com/encyclopedia/term/48751/packet-switching.

"PD-100 PRS – Your Personal Reconnaissance System," *Prox Dynamics*. http://www.proxdynamics.com/products/pd_100_prs/.

"Personal Computer," *PCMAG.com*. http://www.pcmag.com/encyclopedia/term/49133/personal-computer.

"PETMAN – BigDog Gets a Big Brother," *Boston Dynamics*, http://www.bostondynamics.com/robot_petman.html.

"Phantom," *DJI Innovations*. http://www.dji.com/products/compare-phantom.

"Predator B UAS," *General Atomics*, http://www.ga-asi.com/products/aircraft/predator_b.php.

"Predator C Avenger," *General Atomics*, http://www.ga-asi.com/products/aircraft/predator_c.php.

"Predator RQ-1/MQ-1/MQ-9 Reaper – United States of America," *Airforce-technology.com* http://www.airforce-technology.com/Projects/predator-uav/.

"Predator UAS," *General Atomics*, http://www.ga-asi.com/products/aircraft/predator.php.

"Producing, Operating and Supporting a 5th Generation Fighter," *Lockheed Martin*. https://www.f35.com/about/fast-facts/cost.

"Raytheon's Mobile Land-Based Phalanx Weapon System Completes Live-Fire Demonstration," *Raytheon*, December 2, 2010. http://raytheon.mediaroom.com/index.php?s=43&item=1715.

"RC Car to Robot," *Instructables*. http://www.instructables.com/id/RC-Car-to-Robot/.

"Recon Scout Throwbot LE," *Recon Robotics*. http://www.reconrobotics.com/products/recon-scout_throwbot_LE.cfm.

Rodday, Nils, "Hacking a Professional Drone," *RSA Conference 2016*, Feb. 29-March 4, 2016. https://www.rsaconference.com/writable/presentations/file_upload/ht-w03-hacking_a_professional_police_drone.pdf.

"RQ-4 Global Hawk: High-Altitude, Long-Endurance Unmanned Aerial Reconnaissance System," *Northrup Grumman*, http://www.as.northropgrumman.com/products/ghrq4a/assets/HALE_Factsheet.pdf.

"RQ-11 Raven Unmanned Aerial Vehicle," *Army-technology.com*. http://www.army-technology.com/projects/rq11-raven/.

"SkyGrabber," *Sky Software*. http://www.skygrabber.com/en/skygrabber.php.

"Solo," *3D Robotics*. https://store.3drobotics.com/products/solo?_ga=1.253929997.442550321.1451850277.

"Switchblade – Miniature Loitering Weapon," *Defense Update*, 2011. http://defense-update.com/products/s/switchblade_31122010.html.

Szabo, Mark, "Let's hack a drone!," *Github.com*, May 2016. https://github.com/markszabo/drone-hacking.

"Talon Specifications," *Robotinfo.wordpress.com*, http://roboinfo.wordpress.com/2010/04/09/talon-specifications/.

"TCP/IP," *PCMAG.com*. http://www.pcmag.com/encyclopedia/term/52614/tcp-ip.

"The Throwbot XT with Audio Capabilities," *Recon Robotics*. http://www.reconrobotics.com/products/Throwbot_XT_audio.cfm.

"The Totally New SARAH Unmanned Aerial System," *Flying-Cam*, February 7, 2011. http://www.flying-cam.com/en/news.php?id=108&PHPSESSID=9baa6b326c5ae8c672eb2328a62d198c.

"UAS Advanced Development: Raven RQ-11A," *AeroVironment*. http://www.avinc.com/uas/adc/raven/.

"UAS Advanced Development: Switchblade," *AeroVironment*. http://www.avinc.com/uas/adc/switchblade/.

"UAS: RQ-11B Raven," *AeroVironment*. http://www.avinc.com/uas/small_uas/raven/.

"UAS: Wasp AE," *AeroVironment*. http://www.avinc.com/uas/small_uas/waspAE/.

"Unmanned Aerial Vehicles," *GlobalSecurity.org*, http://www.globalsecurity.org/intell/systems/uav-intro.htm.

"U.S. Army increases orders for SwitchBlade microUAV based guided weapons," *Defense Update*, December 5, 2014. http://defense-update.com/20141205_u-s-army-increases-orders-for-switchblade-weapons.html.

US Patent 7339516, "Method to Provide Graphical Representation of Sense Through the Wall (STTW) Targets," issued March 4, 2008, http://www.patentstorm.us/patents/7339516.html.

"Wasp III," *U. S. Air Force*, January 1, 2013. http://www.af.mil/information/factsheets/factsheet.asp?id=10469.

"What is Synthetic-Aperture Radar?," *Sandia National Laboratories*. http://www.sandia.gov/radar/whatis.html.

"World Wide Web," *PCMAG.com*. http://www.pcmag.com/encyclopedia/term/54867/world-wide-web.

Index

Page numbers in *italics* refer to figures, those in **bold** refer to tables.